- Integration: 29
- Leasing with Site Acquisition: 29
- Utilities: Power and Backhaul/Fronthaul 31

What is a Mini Macro? 33
- Why Install Mini Macro Cells? 35

The rise of CRAN and C-RAN/cRAN 36
- What is CRAN and cRAN/C-RAN? 36
  - What is the difference between a small cell and CRAN? 37
  - What are the advantages and disadvantages of small cells? 38
  - What are the advantages and disadvantages of CRAN? 38
  - How do I choose which to deploy? 39
- What about the cost and payback? 40
  - CRAN systems and costs: 40
  - Small Cell systems and costs: 41
  - What's the cost difference? 42
- Would you Deploy Small Cells or CRAN? 44

Would you deploy DAS or Small Cells? 44
- Do you need to choose? 45
- Will DAS and Small Cells Work Together? 46
- DAS and CRAN 47

What G makes sense for small cells? 49
- 2g? Are you kidding me? 49
- 3G was not hot either. 49
- 4G and LTE make Sense. 49
- And now, 5G! 49

Small Cell Opportunities 50
- Overview by Market: 52
  - Enterprise: 52

- Indoor: 53
  - Outdoor: 55
- The real winners: 56
- Why not just put in more macro sites? 57
  - What about the Massive MIMO Macro? 57
- How will 5G that change the small cell model? 59
  - Extreme Broadband 59
  - Ultra-Reliable Low Latency 59
  - Massive IOT Connectivity 60
  - Resources: 61
  - Summary: 62
    - Outdoor small cell vendors: 62
    - Resource 63
- Increase Small Cell Value 64
  - Will Small Cells work with IOT and become the FOG edge? 66
  - Could a small server be put in small cells to control IOT and act as a FOG server? 66
  - Could IOT feed small cell growth? 66
  - Make small cells part of the 5G solution. 66
  - It's all about the Value! 66
  - Resources: 66
- Small Cell Installation Checklist 68
  - Quick, high-level checklist: 68
  - Question Checklist: 69
  - Site Acquisition Checklists 70
- Use Small Cells to Build a Private LTE Network 73
  - Build your own Private LTE Network 73
    - Why Private LTE? 73

Small Cell and CRAN Deployment Report

Wireless Technology Analyst Update

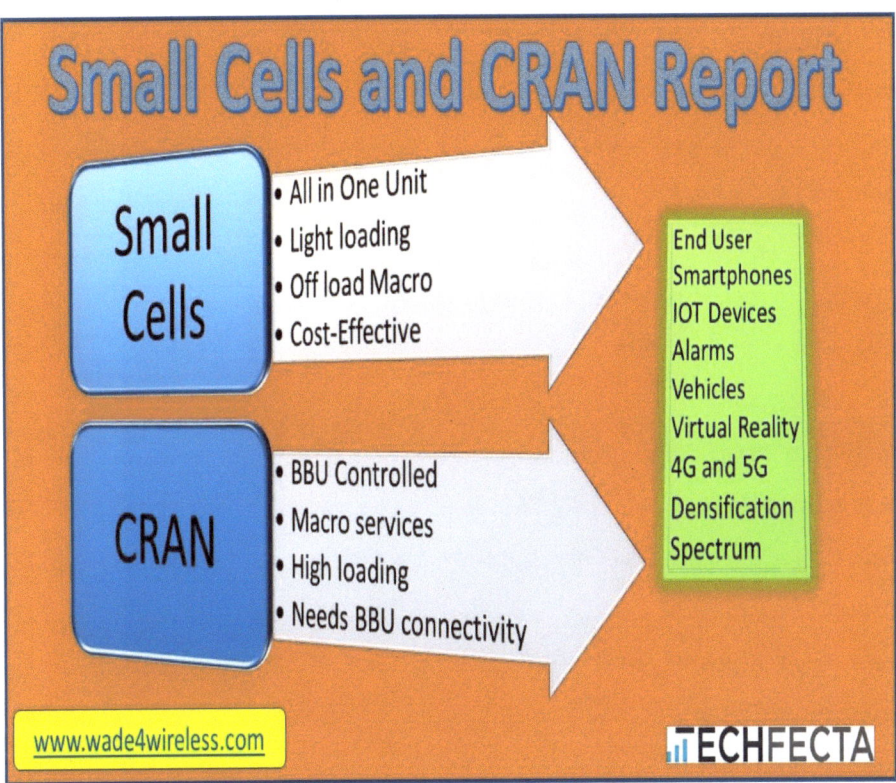

Figure 1 Cover

Small Cell and CRAN Deployment Report

Table of Contents

Contents

Table of Contents 2
Table of Figures 7
Copyright 8
Thank you! 8
Introduction 9
Small Cell History 10
    A quick History Lesson about Carrier Coverage 10
    The business of repeaters: 12
    The Business of DAS: 13
    The Birth of Small Cells 14
What is a Small Cell? 16
    What do People call Small Cells, but they are not: 18
    Resources: 18
Why (or why not) Install Small Cells? 19
    Deployment Options 19
    Indoors (Homes) 20
    Indoor (Business) 20
    Outdoors 22
    Design Flaws 23
    Small Cell Evolution 24
    Deployment models 26
    Top 5 Small Cell Deployment Hurdles 27
        Small Cell Development 27
        Cost of hardware: 28
        Installation: 29

Why would I want a private LTE system?	73
What is the CBRS?	74
CBRS and the Shift in Spectrum Ownership	76
Your Private LTE Network	81
Resources:	81
Small Cells in LAA, CBRS, LTE-U are 5G Building Blocks!	83
Carrier Aggregation and Private LTE	83
LAA – Licensed Assisted Access	83
CBRS – Citizen's Broadband Radio System	84
Carrier Aggregation made this possible!	85
What about the devices?	86
How can this help you?	86
Summary:	88
Resources:	88
Indoor Coverage Matters!	90
How will we cover inside?	90
Wi-Fi	91
LTE-U	91
DAS systems	91
CBRS	92
Inside coverage summary	94
The Common Carrier Small Cell	96
Mounting Small Cells in the City	98
City Deployment Notes	98
Expense Reduction:	98
Now, let's look at ways to make money of existing city asset.	100
City Asset Audits	102
What assets can we mount in a city?	102

- Notes: 108
  - Resources: 109
- Acronyms and Definitions 109
- Thank you Again! 112
- About TechFecta 112
  - *WHAT WE DO:* 112
  - *WITH WORKING KNOWLEDGE OF:* 112
- More Reports and Books: 113

Table of Figures

Figure 1 Cover 1
Figure 2 Passive Repeater 10
Figure 3 Active Analog Repeater 11
Figure 4 DAS Analog System 12
Figure 5 DAS Digital System 12
Figure 6 Non- Pen Mount Picture 34
Figure 7 Outdoor Cabinet Picture 35
Figure 8 Mini Macro Block Diagram 35
Figure 9 Small Cell and CRAN Models 37
Figure 10 Pillars of 5G 61
Figure 11 5G Building Block, CA, LTE-U, CBRS 83
Figure 12 Wade 114
Figure 13 Back Cover 115

Copyright

First Edition © 2018 by Wade Sarver. All rights reserved. No part of this publication may be reproduced, stored in a retrieval system, or transmitted in any form or by any means, electronic, mechanical, photocopying, recording, scanning, or otherwise, except as permitted under Sections 107 or 108 of the 1976 United States Copyright Act, without the prior written permission of the author.

I am not a lawyer or an actively certified safety expert. This book is completed based on research and my experiences. Safety processes and procedures are constantly updated and improved over time. The material contained is for reference only and may include products, information, or services by third parties. I do not assume responsibility for any third-party material referenced in this book.

This document is a guide to help people and not a guarantee that you will do everything properly. By reading this, you agree that myself and my company is not responsible for the success or failure of your business decisions relating to the information presented in this guide.

www.wade4wireless.com

www.techfecta.com

Cover and design by Wade Sarver

Thank you!

Thank you for purchasing this report, I appreciate your support. I pray that it serves you well.

Your feedback is needed, let me know what you like and didn't like about this report. What should I add, and what should I skip next time.

If you need one on one consulting or specific reports, feel free to reach out at wade@techfecta.com for direct support.

Small Cell and CRAN Deployment Report

Introduction

This report is to help you understand more about small cell and CRAN deployments and all associated options. When you work in the industry people, commonly lump other things into small cells or small cells into other things. For instance, CRAN and cRAN centralized RAN and Cloud RAN often looked at as small cells because a radio head is being deployed on a pole or building all by itself. That is just like a real small cell; it's installed all by itself.

Another thing is the mini macro; it's just a big and power small cell. It has the form factor of a small cell but can do more. Not quite as much as a macro site, but more than your typical 1-watt small cell.

Then there is DAS. DAS incorporates small cells, Wi-Fi, radio heads all into its system for the carriers. It could be one of these or all of these. DAS systems for 4G and 5G are going to be all digital.

It's going to help you look at small cell deployment holistically. There are deployment notes, history, and an outline of what works and doesn't work. This report covers more than small cells to give you a big picture of the future of wireless outside of the macro site.

Then comes the ever-important deployment. You know, on paper it all looks good, but when the deployment happens, you must get it out into the real world and make it work. That's the magic. All the planning in the world can't overcome problems that you run into. Only trial and error.

These notes in this report are here to help you understand what you may run into and how to overcome some issue. The history is to help you understand why things are done the way they are done now.

It helps to understand what the carriers are doing now to make the deployment evolution clear. As we go to an all-digital system for wireless communication, devices become different. They must pass data. If you want voice, instant messages, Facebook, Google, or whatever, it's in the data as a data packet. The radio doesn't care; it must pass data. Just like the backhaul, it doesn't care; it must pass data.

Small Cell and CRAN Deployment Report

Small Cell History

Small cells have been the buzz around the industry since about 2013. We thought that they would be the new site deployment. I believe that all the carriers wanted to believe it as much as I did. You see, I was working for Alcatel-Lucent and we had every intention of rolling them out to the industry by the 10s of thousands! As you know, that didn't pan out, and I will tell you why.

First, let's take a stroll down memory lane and see how small cells were developed and why.

A quick History Lesson about Carrier Coverage

The reason small cells came about is that in a digital world the solutions of past would not serve us as well going forward as they did in the past. The spectrum and technology that we had in 2G were not helping us in 3G and 4G would make sure it's out of reach.

We worked with passive repeaters, which consisted of an antenna at each end and a coax cable in between. The design would be that the gain of the antennas would overcome the loss in the cable and take the cable from point A to point B.

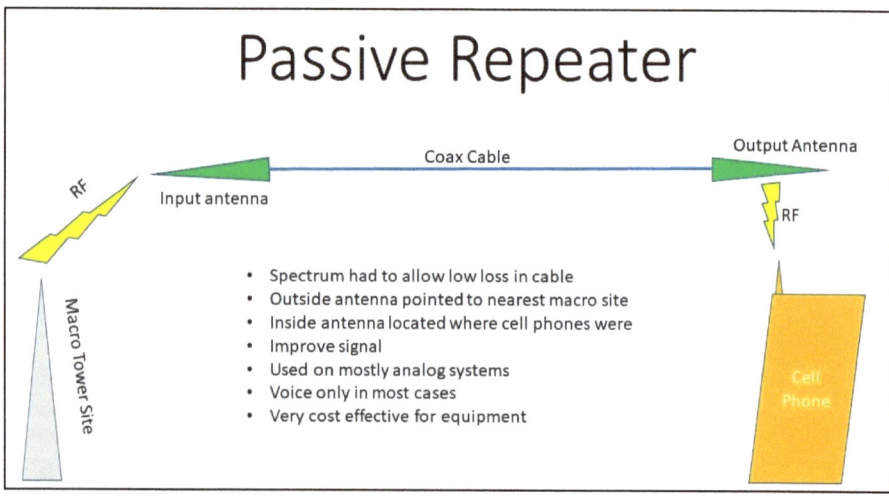

Figure 2 Passive Repeater

This was fine until 3G came along, then it didn't work as well. Then as data usage climbed, it became more of a problem with data. There were other systems where they had an amplifier in between the 2 antennas. This would take the input signal and amplify it so that the output signal we very strong.

Now, don't' get me wrong, these repeaters were used for a long time, and they worked. They solved the coverage problem that so many businesses had. Back then the carriers relied on rooftops and tower for over 95% of coverage. Even then they would not pay for smaller systems. If someone wanted coverage int heir building, he or she had to do it themselves. Repeater systems were normally under $1,000 each and installation could be a few hundred dollars depending on the complexity. You had to be a big customer to get the carrier to chip in on the cost.

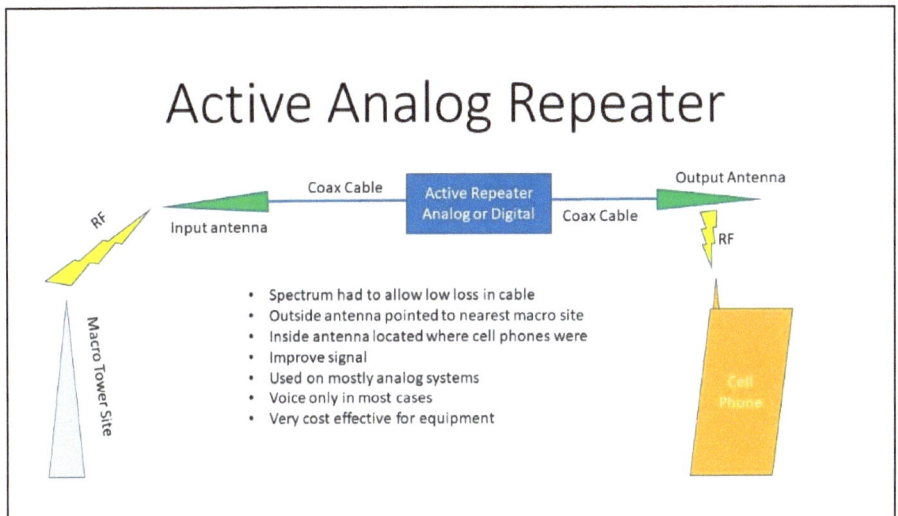

Figure 3 Active Analog Repeater

The repeater could have been analog or digital, and you could have multiple antennas. The idea was to get the signal to the mobile. It worked well. Especially in 2G and 3G, it was reliable.

This type of architecture eventually became a stepping stone to DAS systems. The economics behind DAS were to get the signal to a concentrated area. The carriers didn't want to take this on early, so they would pay people to build out the DAS systems and install them.

First, here are some models of DAS systems, granted, I draw like a 10-year-old, but you get the idea.

Small Cell and CRAN Deployment Report

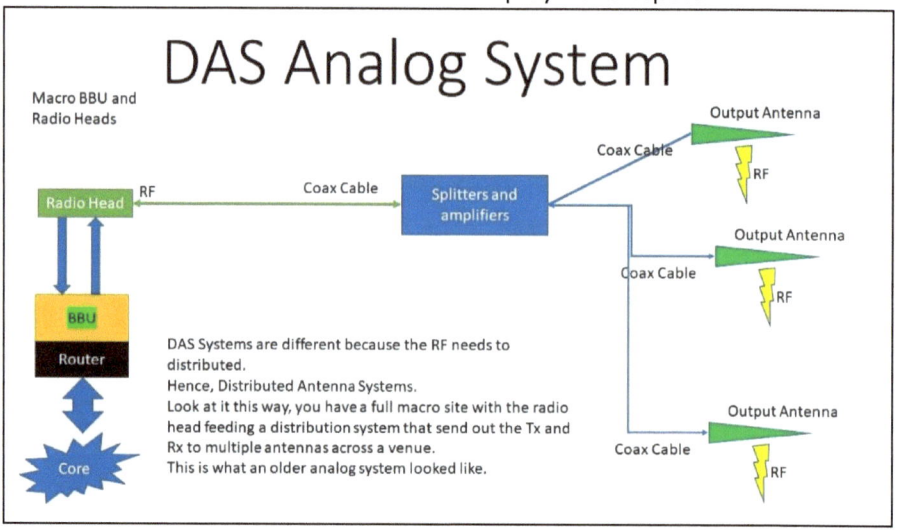

Figure 4 DAS Analog System

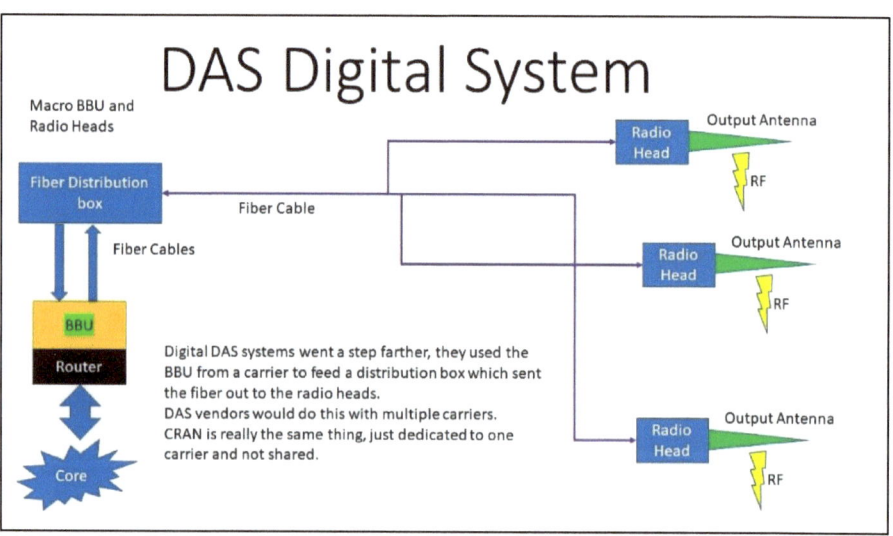

Figure 5 DAS Digital System

The business of repeaters:

When repeaters or cell extenders or whatever you want to call them came about, the carrier would not do it unless it was a larger customer. Remember this was in the 80s, 90s, and early 00s. Usually, the customers had to pay for themselves. That means the business had to find a local radio shop who would design it and install it. They didn't always work so well, but if the signal was better inside the customer will be happy.

The carriers didn't like putting them in because of the expense, so unless the customer had 50 or more phones, it really wasn't worth it. However, customers began to rely on their mobile phones more and more, so they would pay to have coverage everywhere, so they would not miss a call.

I remember when Nextel had the push to talk, PTT, feature. This made it critical for coverage because contractors used this feature all the time. It became a lifeline for them to do business. So much so that the phone call was the second alternative.

The Business of DAS:

Then came DAS systems. While these were not cost effective for a smaller business, they solved many problems for the carriers at larger venues. The DAS systems would cover a venue quite well if designed properly. DAS was around for 2G, 3G, 4G, and will continue into 5G. DAS is still used in public safety because many municipalities require a usable signal in a building. So, for a new building to be open to the public, there should be some type of public safety DAS installed. However, in today's world, the carrier DAS is not so popular unless it is absolutely needed for the carriers' customers.

What venues would need DAS?

- Airports
- Sports Stadiums
- Concert Halls
- Convention Center
- Large corporate parks
- Industrial parks
- Large office building common areas
- Shopping malls
- Hospitals
- Train and bus stations
- Businesses
- Industrial buildings and complexes
- Large office buildings
- And the list goes on.

There were a lot of places where DAS systems could be installed. The business models early on were somewhat simple. The carriers were putting them in, but soon found maintaining them was a headache. So, then they let other put them in and paid rent to be on them. Eventually, the DAS installers thought they would get more money if they charged per call or connection, and the carrier thought they would save money, either way, managing that system was a complete nightmare, and the numbers were impossible to trace.

Then the digital systems were put in, and most DAS installers went back to the flat fee system. While this may have been good for the landlord, the carriers were tired of dealing with it. They decided there had to be a better way for smaller venues. Let's face it, DAS ownership was very profitable at one time, so the building owners wanted in on it, and the DAS managers were being squeezed.

The carriers decided to support larger venues; other customers were on their own. So, the groups that leased DAS systems had to find new customers. Guess what, the landlord didn't want to pay to have a DAS system, but they wanted paid to have it in their buildings. The fight went on until someone needed coverage more, either the landlord would make his tenants happy, or the carrier would try to satisfy the customers in the building. Newsflash! The carrier really didn't see any need to satisfy anyone in almost all buildings. The landlord had to decide, who would pay? I'll give you a hint; the landlord won't pay unless he has to like for public safety. A carrier would need to see payback either in a large venue or a large customer before spending the money to install DAS.

The Birth of Small Cells

There were Wi-Fi hotspots, why couldn't the carriers do the same? Well, they did. Small cells came about because carriers wanted to emulate Wi-Fi hotspots, and the dream was to put them everywhere a macro site could not be for a lot less money. This sounded great! The hardware was cheaper, the energy usage was much less, and they could bypass the DAS systems altogether.

The idea for a small radio was around for some time. Often small repeaters were used, but as 3G progressed, the idea was to have indoor coverage, period. If small cells could be deployed like a hotspot, then the carrier could put them everywhere. Ah, the dream of seamless coverage, but that didn't quite pan out.

Carriers were not happy about connecting small units to the core through a VPN line that went to the open internet. This was the most cost-effective way to do it, but they had security issues. This was a problem early on. The 3G systems also were a challenge to get that small. There were a lot of issues that could not be resolved before 2010. In the days of 3G, a small cell was used for coverage more than loading. Occasionally they would relieve loading, but not normally. Let's look at this in more detail.

Small Cell and CRAN Deployment Report

What is a Small Cell?

OK, what the hell is a small cell? Is it just a small cell site in a box?

I guess the first thing we need to do is outline what a cell site consists of. Picture this as I walk you through the site starting at the antennas. The antennas, normally on a tower in a Macro setting, would be up in the air. With LTE they put the RRH, (that is an RF unit that normally transmits over 30 watts), up on the tower close to the antenna. The RRH and the antenna are connected using coax line, (RF cable). For the sake of this book, I am going to concentrate on LTE. What is coax? That is a cable that can carry RF signal. The other connection to the RRH is fiber that runs to the BBU. The RRH is powered by a power cable. So, in a Macro setting, they must run the fiber and power up the tower to the radio head. If we follow the fiber and power line down the tower, chances are good it will run into a shelter or cabinet to the controller. The controller is called a baseband unit (BBU) which will have all the information in it. The BBU will get its information from a router at the site, a cell site router. The BBU at a Macro site could control 1 to 12 RRHs. Normally we call the BBU and the RRH a BTS (Base Transmitter Unit) or an eNodeB.

So back to the question, what is a small cell? It is the BBU, and the RRH all crammed into one unit, an easy to mount box. It may have the antenna attached internally, or it may have coax, (pronounced Kō-Ax), jumpers connecting an external antenna. It is usually very small and can be mounted anywhere. The small cell will have fiber connecting to the backhaul which will connect it to the EPC.

It's like putting a small low power cell site where the people are; I mean right where the people are using the most data instead of way out at a tower site or up on top of a high-rise building. Just picture an eNodeB, (the transmitter for a carrier). This can also be called a BTS, (base transceiver station) which is located at the site. It receives and transmits the RF signal. Suddenly you have a cell that can service a group of people with little RF loss and high data throughput. This is a great thing for carriers because it can fill a small dead spot or improve offloading in a smaller area where the Macro site can't. That is the goal of the small cell, to alleviate the loading and add coverage. I will concentrate on the outdoor small cells for this article, but the indoor do the same thing.

Usually, a small cell is 5 watts or less. Sprint is creating the Mini Macro, a small 20-watt cell. This is another small deployment that could be put on a utility pole. The outdoor 5W small cell is getting smaller, about 10 lbs. or less and usually less than 18 inches by 18 inches by 10 inches. Low power and a small profile.

What about the backhaul? You may need a router if one is not built in. Many small cells now have the router build in, but if they don't, it will be something you need to have at the site. The backhaul could be fiber, copper, or wireless.

You may think that it's something anyone could put in, right? Well, the indoor small cells are easier to install, but outdoors you still have many issues. Although depending on who the Original Equipment Manufacturer, (OEM) is then you may need their training to be authorized to install it. After all, if it is a high-tech piece of equipment that a carrier trusts you to install properly. They probably want you trained to install it, so the warranty is not void. I will go into that later.

The small cell is going to change the way we look at wireless deployments. The biggest band for your buck now may be to go where the people are, not to be as high as you can possibly be. Also, think of the spectral efficiency by having smaller sites, that spectrum which costs billions is now used more efficiently. The smaller the site, the more you get for the same spectrum. The cost of spectrum and the need for maximum bandwidth is changing the way cell sites are deployed.

Small Cell and CRAN Deployment Report

A small cell is a stand-alone small cell site, usually at very low power. It will have an integrated BBU and backhaul. It may have a router inside but, in most cases, it will connect to an external router, switch, or hub. It is a standalone unit that needs backhaul and a way to connect to the core or a controller at the core. Generally, this unit could be put anywhere and is built for indoor or outdoor use.

What do People call Small Cells, but they are not:

Here's the thing, most deployments that installers think are small cells, really aren't technically a small cell. A true small cell is a self-sufficient low power BTS site. It has backhaul which in some way connects to the core. Most installations are either a CRAN or cRAN which is a remote radio head that connects to a BBU located somewhere else, or a mini macro which is just like a small cell but higher power and has more functionality. More on these technologies coming up.

Resources:
- https://www.thinksmallcell.com/System/what-is-a-small-cell-or-femtocell.html
- https://www.thinksmallcell.com/Technology/a-quick-recap-on-new-and-upcoming-small-cell-technologies.html?awt_l=7P1V2&awt_m=J7bb0bboD_CA6G
- https://wade4wireless.com/2017/03/06/mounting-small-cells-in-a-city/
- https://wade4wireless.com/2016/10/11/deploying-5g-small-cells/
- https://wade4wireless.com/2018/03/26/increase-small-cell-value/
- https://wade4wireless.com/2016/05/24/art-king-is-the-king-of-enterprise-small-cells/
- https://wade4wireless.com/2016/01/04/small-cell-installation-checklist/
- https://wade4wireless.com/2015/11/19/free-the-lte-indoor-small-cells/
- https://wade4wireless.com/2015/07/27/small-cell-fronthaul-and-odas/
- https://wade4wireless.com/2015/11/12/wireless-deployment-handbook-for-lte-small-cells-and-das/

Why (or why not) Install Small Cells?

I will tell you why, coverage and offload. That is why we all put Wi-Fi in our homes, so we have broadband everywhere, right. Well, the small cell solves that problem with our wireless devices. We want to have good coverage, and we want great broadband speed, so we can upload and download our important apps, like Facebook or LinkedIn. You know the stuff that millions of people rely on all day every day. Not to mention email! I was going to mention YouTube because that is a major app for mobile devices in 2015.

The coverage issue has been around for a long time, decades, but now we have an opportunity to do something about it by installing small cells. I know that many of you really like Wi-Fi, but the issue here is who owns the Wi-Fi and how can use it. Also, interference has been an issue for a long time.

Wi-Fi is awesome and saves you money on your mobile phone bill. Voice over Wi-Fi, (VoWiFi), works well if you have a decent backhaul. But in a shared environment, you will have issues. I know people who have made calls on VoWiFi. In a dedicated Wi-Fi the calls are clean but, in an airport, they cut in and out. I love Wi-Fi, but I also know the limitations in the real world can be painful if you rely on your device. So, the carriers are there to make sure you have access to their system, for a fee of course.

However, this is changing because carriers are looking to the unlicensed bands to offload voice as well. I put in more about this in upcoming chapters. The carriers are working to provide aggregation. Read on to learn more.

Deployment Options

There are many deployment options. Let's look at the indoor and outdoor options. The one thing about small cells is that they connect to the carrier's core. When the CBRS and other non-carrier options take off, then we can explore new areas to deploy small cells outside of the carrier realm.

The thing is, now we see any small cell would connect to a core of a carrier. Even the Femtocells that the carriers send to people's homes connect to a server in the carrier's core. T-Mobile has done this with great success. They could supply people with a Femtocell that provided excellent licensed and unlicensed coverage on one's home. Was it worth it? Only T-Mobile can tell you that. They must weigh out the cost of not only the unit, but the setup and initial work to create something that could be plug and play in someone's home through an independent internet connection with unknown firewalls that could block out the connection.

Let's concentrate on the industrial uses here.

Indoors (Homes)

The home small cell market is a different niche that is geared towards the consumer, completely plug and play for the end user with no installation or planning necessary, hypothetically. It's a different market, that's the point.

For home use, the femtocell was born. Carriers would not pay for it, but they would sell it to users. AT&T built up a nice business selling it to homes in suburban and rural areas. People were willing to pay for a device to make their coverage in the home as good as outside.

Of course, 4G was all about data, and the Wi-Fi router made the data work well in the home. Why pay for anything more unless you relied on your device for calls, which was becoming less and less important after 2016.

Indoor (Business)

Inside buildings coverage was nice, but small cells were not as cheap as the carriers hoped. They were cheap hardware and installation which made it cost effective, except for that one thing, backhaul!

Yes, backhaul, that nagging problem that faced us all. Once again, the carriers had to foot the bill to add a T1 or some way to get the connection back to the core. Remember back then T1s were used for much of the transport. Today we rely on all data networks, which has made things easier in some ways, but we have new problems. The carriers had to manage these units. They had to run them up.

In the home, with people who had internet access, the units became plug and play and would connect seamlessly. The femtocell became the go-to device for home usage. Again, the end user would have to pay for it or, someone like T-Mobile would make it free to end users. WOW! This was a gamechanger for the business that Verizon and AT&T may have to compete with. Although, it still isn't quite as mainstream as we had all hoped.

Then, the small business tried something similar. It is the same model only they call it a Pico cell which does the same thing.

Small Cell and CRAN Deployment Report

By now, the femto and Pico cells are almost all plug and play, and they are so easy to install. They are almost all becoming 4G units that are mostly data with some voice over LTE, VoLTE.

This is where the digital systems look so promising. Suddenly they could VPN through an internet connection to the core. They could use the customer's internet to make this happen.

The VPN came with a lot of challenges, like latency, secure connections, fear of being hacked, loading up the neighbor lists, have control of all the small cells, loading the golden parameters for each cell.

Suddenly, the small cell became as much work to integrate and setup as the macro site, almost.

While indoor small cells were a better deal, but they had limitations. They were limited to carriers and throughput and loading. This isn't such a big deal for small offices, but they can't handle the larger venues well at all, they get overloaded quickly.

They were starting to roll out in larger venues, like train and bus stations, lobbies of business buildings and airports. However, DAS was still working better for larger venues. It could handle the traffic much better.

The indoor small cell market did take off. It really started to boom with the growth of 3G and 4G. it will continue to grow even more with 5G. What really helped it was how easy the small cells were to install and the ease of installation. They overcame the backhaul hurdle and the integration hurdle. Backhaul through the customer's internet access and integration is done by plug and play by connecting to a remote server that would connect it to the proper core. WOW!

One more thing, vendors of small cells grew beyond the traditional OEMs. Suddenly we had companies beyond Alcatel-Lucent, Nokia, Samsung, Ericsson, and Huawei. We have SpiderCloud, who did an outstanding job getting certified on carriers' systems and marketing to enterprise users! We have Airspan who also made a dent in the small cell world along with ip.access, Airvana (purchased by CommScope), NEC, ZTE, Parallel Wireless, and more.

The growth of indoor small cells looks promising because most businesses want seamless coverage and are willing to pay for it if the product and installation are reasonable. They prefer no OpEx, which is how most models are set up. The carriers win because they have a unit that is almost completely plug and play and they get additional coverage.

Outdoors

Now, let's go outdoors. The outdoor model was everything that the carriers wanted, cheap hardware, small enough to mount anywhere, cheap for installation. They got what they wanted. That's why carriers should not design their own small cell. They overlooked the site work, like getting power to the site, backhaul to the site, and how loading would change.

However, as the carriers soon learned, the hardware was the least of the cost. There are still the installation costs that go with it. While the actual installation was reasonable, the backhaul was very high, note efficient as all for such a small site.

Then the site acquisition and documentation also cost a lot of money, not to mention the rent could be very high for the small area covered. The outdoor small cell suddenly became a burden. It was only deployed where it had to be deployed.

Remember that when small cells came out in mass production, there were 3 technologies. You needed 2G, 3G, and 4G all in one tiny box. They could be spread across 3 or more bands, all in one small box. This was not practical in any way. There had to be a payback for the carrier to deploy these in one area. Voice generally wasn't enough cause to deploy a small cell.

Welcome to the world of digital, 4G specifically. Now that 4G LTE is everywhere, and the carriers can do voice over 4G like is easier for small cells. The backhaul is all digital, usually fiber, but the small cell itself is more like a Wi-Fi hotspot with a beefed up backhaul. Generally, with one band only. It's still not near a carrier's performance, but it's good enough to make it a cost-effective fill site.

With the growth of 5G outdoor small cells will be able to do more, but it's the backhaul making it a less attractive option. Wireless backhaul makes it a better option. Specifically, what Sprint has been doing in 2018 with the 2.5GHz spectrum. They found a way to make a small box with enough licensed backhaul to use as a fill site. Great idea!

The thing is, small cells outdoors should serve a higher purpose and perform very well. It's not just about filing a dead zone but improving local performance in a specific area or offloading an overloaded macro. If an area is a dead zone, this is a way to provide coverage.

Outdoor small cell deployment did not take off, yet. The payback for the carrier wasn't there. Now, with 5G, it is expected to play a bigger part. With 5G, there are expectations that throughput and reliability will be better than ever before, so the small cells or remote radio heads (CRAN)will be needed to make that happen.

Design Flaws

One thing I learned from working for an OEM is that you should not let carriers design small cells. They think they know what to do, but they may not understand why all the features are added there.

In a post, I talk about how a carrier thought they knew what they wanted, but it turned out to be a disaster. Here's a story of how things can go bad quickly.

A carrier wanted to design their own indoor small cell. It was interesting because they went beyond just telling the OEM what features they wanted, but they wanted a hand in the complete design. They said they would commit to over 1,000 units, but they were sure that sales would be 20 times that number. So, they started outlining what they wanted.

Typically, we would have a model and add the features a carrier wanted but, in this case, they knew everything and would only commit to it if they could get what they wanted, not the model or features the OEM would recommend because they felt the OEM only wanted to sell an expensive product and did not have the carrier's best interest in mind. Talk about paranoid, or maybe just cheap.

The cost was a factor, so, to save money they said they didn't want PoE, (Power over Ethernet) because they wanted to cut costs on the device, (among other things but let's go with this since it plays into my story). From that perspective, they saved a little money. Unfortunately, the carrier didn't think beyond the device cost. Once it came to installation, they had to get power to every location that the small cell would be mounted in the office space. This was a small unit with no PoE, meaning they either had to run power cords along with CAT 5 to every location or they had to install outlets at every mounting location in the ceiling. See the problem? Think about mounting a small device in a ceiling and having to run power to every device! It cost a fortune to run electric outlets everywhere, and extension cords are ugly, and there are regulations with running AC power in a ceiling. In a ceiling, AC lines must be fire rated or in conduit or maybe both. Think about it, if someone cuts a ceiling tile and there is AC extension cord laying on it, ZAP! We have a problem. These problems happened because a carrier's team thought they would save a few dollars on hardware, they failed to see the big picture. Sure, we tried to tell them, but they just stood firm, convinced we wanted more money out of them. Everyone's investment money wasted! The small cell model was scrapped after about a year, and the OEM was lucky that the carrier accepted delivery on a few hundred, they didn't honor the full commitment.

This is common when a group does not look at the big picture; they often are very short-sighted.

In the beginning, small cells covered what the carriers wanted, cheap hardware and easy to install. Once again, the carriers didn't look at the end to end installation. They didn't think about site acquisition, or backhaul, or permitting issues. What the hell were they thinking? They learned quickly about the other issues. Also, they demanded that the units were plug and play, well now they are, and it didn't cut the backhaul costs in any way shape or form. In fact, it caused new problems; someone had to configure the router, test the backhaul connection, and verify the bandwidth.

Small Cell Evolution

The small cells have evolved over the last 5 years. Originally the carriers wanted an all in one unit, meaning antenna and everything, like your Wi-Fi router. This we a problem because the antenna could have been directional or omni, but it limited the design.

Spectrum is an issue; it's not a Wi-Fi unit where it's very low power, and you just blast everything. Usually, in a home or small business, there is little planning done for Wi-Fi, they just throw it in and add units as needed. Professional Wi-Fi systems are planned and designed and have the channels assigned based on coverage and loading. However, in most situations, people don't spend the money unless they are the only Wi-Fi supplier and quality of service, QOS, matters.

You only have one technology, Wi-Fi. Sure, it could be 802.11ac or some other version, but they are all backward compatible. They all talk to each other if it's not too old.

With a licensed small cell, you could be dealing with multiple technologies, 3G, 4G, and now 5G. If you look at this, 3G could include GSM or CDMA. Then 4G is pretty much all LTE, WiMAX is not popular anymore, and no major carriers are not using it in the USA. The reason technology was such an issue was not only because of the spectrum, but it was more work for the small cell.

Now, with LTE being the dominant format, 4G, things are getting easier for all of those that have VoLTE and can-do data and voice over the same 4G carrier.

Let's talk about spectrum. Another thing the carrier must deal with is the multiple bands of spectrum that is out there. For most small cells they try to settle on one band. They would like to have multiple carriers in the band, if available, to get more bandwidth. Not all small cells can process multiple carriers.

Also, the planning between the indoor system and the outdoor system. The small cells need to know whom they will hand off to. The calls have to go somewhere, and the small cells need to know what it's options are. There are golden cells that allow the UE device to handoff from the indoor system to the outdoor system. They are treated as different systems. The small cell tells the device where to go, literally. The UE will go where it is told, and if it can't connect to the other cell, the call drops. Too bad, dropped a call, unhappy customer, complaints roll in. All because a setting was wrong, or the configuration had a bad neighbor list. With a cell system, planning and design are not luxuries but necessities. Small cells are part of that cell system.

Outdoor small cells became better and better. One of the main problems was the power setting; if it's close to people, like about 10 feet off the ground, the power should be lower and acceptable per the FCC limits. If it's 20 feet up, then it can be a higher power. One more design consideration.

Outdoor small cells had to be made quiet, mainly if they were put in residential neighborhoods so that people didn't complain. Fans make noise; they had to be creative with airflow. Also, fans are usually the first thing to fail, another problem is solved when fans are removed, and airflow is naturally flowing. Not to mention less electric usage.

To design an outdoor small cell, they put almost the same parameters as a macro cell, just on a smaller level. It must know whom it can handoff to, the locations of each, and how to talk to the UE device. Let's not forget that spectrum is still an issue, along with loading. More factors to consider.

Many of these issues have been resolved. The one big difference between a macro site and a small cell site is the loading. Small cells still have serious loading limitations, so the macro site may still be needed. It all goes back to the design.

Another side story, I was working with a carrier that thought it could use mini-macros and small cells to replace the macro sites or expand using only small cells or mini-macros. Guess what, they didn't plan on the loading, and they put them in any way because the installation and commissioning were so cheap. QOE is going to suffer when you limit the number of carriers and aggregation. These are things that the Guess what, less than a year later they started replacing all of them. They had to put macro sites in because they underestimated how heavy the loading would be. You see, all those features, like carrier aggregation and MIMO, play a huge part in making the site reliable to the end user. Planning is important, and so is design! Let's face; some good old common sense would have paid off here.

Deployment models

I think that people see a small cell as just a simple fill solution. That is not true; each deployment scenario should be looked at for its unique need, not as a generic fill BTS.

Urban – when deploying is urban settings you are probably solving a loading problem. There could be a coverage issue, but in most cases, it's to offload the macro site. While the small cell can't replace a macro site, the small cell helps offload local traffic. It's a good idea to plan to offload so that the macro site can hand off to the small cell. Also, there may be multiple small cells set up to look like a sector facing in a specific direction. Anything to offload the macro site.

Suburban – we need to look at the suburban sites as potentially offloading the macro, but in many cases, it will be to improve coverage. It's hard to say that suburban areas need to offload except in an area where people congregate. Even then, it may be more for coverage and less for loading.

Rural – let's face it, rural, it's coverage. There's no incentive for the carrier to worry about loading in rural areas. Nuff said!

Enterprise – this is up to the local company. If they use their phones for voice, who cares about loading. They usually have Wi-Fi but may rely on LTE, but not likely. Chances are good they have Wi-Fi and only rely on the LTE for data backup. Why would you pay extra data when you have Wi-Fi for free?

Top 5 Small Cell Deployment Hurdles
Small Cell Development

People often ask me about small cells, mainly because I was a fan and an advocate of the small cell movement. That's right. I thought small cells were going to be a bomb that exploded and spread across the USA like a cat video on Facebook. I was sure that it had the makings of a significant movement that would enhance the amazing coverage of macrocells in a tidal wave.

Well, I was wrong. The movement had problems that the OEMs and the carriers could not iron out. Want to know what they were? Of course, you do, it still plagues the industry today.

The small cell had 3 of the 5 things that a cell needs to roll out reduced dramatically. The carriers were severely misguided when they put all the pressure on the OEMs and the installer. You see, the carriers, at least here in the USA are very short-sighted when they look at the bills. They were convinced that the OEMs were ripping them off. To this point, they pressured them to create small cells. They looked at the Wi-Fi access point and thought, "Why can't we do that with an LTE cell?". I'll tell you why if you read on oh faithful one.

What are those 5 things? Here's the list

1) **Cheaper equipment** – small cell hardware is very cheap, problem solved.
2) **Cheap installation** – small cells are very cost effective to install, problem solved.
3) **Easy integration** – most small cells today are plug and play if they can talk to the core, problem solved.
4) **Site acquisition and rent** – only half of this problem is solved. Site acquisition is still almost as much like a tower site, around $10K in some cases. The rent issue has recently been resolved in some states with the

Small Cell and CRAN Deployment Report

wireless associations lobbying all the states for cheap rent and reduced permitting. Most states are putting rules in place to limit the amount of money a municipality can charge a carrier. Still a problem.

5) **Utilities** – here is another problem yet to be completely resolved.
   a. **Backhaul** - Fiber can be very expensive to run to a site, almost cost prohibitive. It has not been easy. Sites should be chosen that are easy to get fiber without trenching or expensive rights to poles. Then there are the monthly payments for finer access, not always cheap depending on the provider and pole rent and so on. If trenching is involved, it's expensive. Wireless is just now catching up, but it seems like the best solution when it works. It's not the size of wireless antennas anymore; it's the bandwidth they can provide. You need around 1Gbps, at least 100Mbps, for most small cells to perform adequately but that's not always easy for wireless to provide to a pole. We are almost there with the new mmwave systems, so this may be resolved sooner than later. Still a problem.
   b. **Power** – still an issue is getting dedicated power to a specific location for a reasonable cost. You either have to share power or install your own for a premium. Still a problem.

When looking at the big picture, you can see why the small cell rollout never hit the massive numbers everyone expected back in 2012. It may happen someday, but the need and technology and spectrum had to get there to make it a need. Not only that but each carrier has its own strategy. Some deploy CRAN because the radio head connects to a BBU giving it all the functionality of a macro sector. Some deploy true small cells which are limited in capability but cost-effective. It all depends what their specific strategy will be over the long term. Some install a small cell to see if it will be utilized, then if usage is heavy, they upgrade. That makes sense but can be a drain on resources and finances over time.

Cost of hardware:

The OEMs came out with a very cost-effective piece of hardware that solves the issue of blowing the CapEx budget. They made the units lower than $2,000 US if you order in quantity depending on what you get. I am just saying that this was the focus of the carrier and it was resolved. The carriers were very short-sighted focusing on hardware. They looked at the OEM as the bad guy. How did that work out? You tell me.

Small Cell and CRAN Deployment Report

Cost of hardware is very cheap for what you get, especially if you buy in bulk. Problem solved!

Installation:

The installation of the small cell went way down, almost cheap. There are still grounding issues, heat issues, noise issues, and grounding. The antenna azimuth and tilt still matter. The installation is very cheap, but there are still stringent standards for mounting the cell, the antenna, and grounding. It's critical. However, installation is way cheaper, so this issue has been resolved!

Installation is cheap. Problem solved!

Integration:

The small cell must be integrated into the existing cluster whereas most Wi-Fi units are just powered up and handed an ESSID and a channel plan, which the channel plan is often optional. Then voila, a live Wi-Fi hotspot. One that anyone can access assuming they get the ESSID and can join. Sure, there may be more to it but not much.

Oh, the joys of Wi-Fi, the device can access it so quickly, then you can sign in using some type of server, and you're in and can roam from access point to access point somewhat seamlessly.

So why isn't the small cell that easy? Because of the handoffs, each cell must know all of its neighbors so that there can be seamless voice or data handoffs. It must know who it is around, so it can handoff to the right cell seamlessly. While the carriers even figured out how to make them plug and play, they still must build a template and get the GPS data from the small cell so that they know where it is and see if the US gets the carrier without any hiccups.

However, plug and play made integration very cheap, problem solved!

Leasing with Site Acquisition:

That nasty rent check that the carriers must pay every month will not go away, no matter how much the carriers try to make you think they are doing you a favor by putting in a small cell to help you get a better signal. You see, the carriers didn't just drop in small cells anywhere, there had to be a need and it had to be cost-effective. They also wanted to save money on rent, which is much cheaper than a macro, but you're only getting about $1/60^{th}$ of the coverage of a macro site.

Small Cell and CRAN Deployment Report

Leasing has proved to be a challenge because it is way cheaper than the carriers could have hoped for, but when you add it up, it's not much cheaper than the macro cell if you look at the coverage comparison. How did this fly with the carriers? You tell me after you read the next statement.

The carriers got the WIA and the CTIA to go around and lobby the FCC to pass a law to allow them to mount anywhere for a little bit of money. That didn't work because of jurisdiction issues. They didn't give up. You have to admire those tenacious wireless carrier advocates. They went to lobby the states. You see, this was easier than dealing with detailed site acquisition for each municipality. It was better to get the states to mandate that the small cells be installed anywhere the carriers want for very little money and with as little interference from the local city or municipality. So far it seems to be working.

I feel the pain of the carriers. The cities and municipalities started requiring permits for everything. They want to know about rooftops towers 10 miles outside of town, and anything mounted on any pole throughout town. It's like when you need to get a building permit to hang drywall in your basement. Chances are you won't do it because you know they won't catch you, but anything outside the local authorities can see it, and they will order you to remove it or pay for the permit and maybe a property tax along with it. You see, the cities and municipalities created this mess so that they could get permit money for every attachment, not they are putting up a fight because they know it's gotten out of control.

However, I don't want to be all "one-sided" on this issue. I have seen some horrible and noisy installations in the middle of cities. I think the installer and carriers would be embarrassed to have such monstrosities out in public. Most don't care, trust me, people call and complain. The city usually has deaf ears unless they have an ordinance already in place. While you may not care, just picture an ugly noisy pole in your front yard. If you look out the front door or open a window, then you must deal with this, and it won't go away. There's the issue, the people that live there learn to hate them and build a force to make sure it doesn't happen again. If the carrier and installers would work to create peace in the community, things may not be this crazy.

My point is, the cities and municipalities are not all at fault. They had to deal with carrier's ugly boxes and antennas on poles that just look awful. They also had to deal with citizens complaining about the noisy fans that cause even more noise pollution in some quiet neighborhoods at night. No one wants to hear fans running all night right outside their home.

So, we need balance. I am not sure it will happen. The pendulum will swing in favor of one or the other depending on how the states rule. California went against the carriers, but many other states will let the carriers roll over them costing the cities hundreds of thousands of dollars to support. The municipalities will do their best to make it back in permitting fees. I am sure they will skyrocket in some cases. You see, most states will say they care about the cities, do they? They want 5G to be rolling out ASAP. One thing I learned, elected government officials love lobbyists from large companies. If they have their support, they may get reelected. You can figure out the rest.

Leasing is more than just agreeing to terms and signing the papers. There is a complete site acquisition process that should be followed for the installation to happen. The carrier needs to do a bunch of tasks like site survey, loading studies, RF studies, construction drawings, leasing costs, legal fees, permitting, and more. So, any site for a carrier involves a lot of work before the BTS, or small cell ever gets installed. It takes planning and coordination to make sure that the lease and documentation are complete. If there is one thing missing, then they may have to start over. Someone must pay; it makes sense to pay only once.

Again, site acquisition can take months; most people have no idea how long it takes. Many times, it could take 6 months to a year, all because of the process. Oh, and it's not cheap, it is so expensive, even for a small cell, regardless of what you do, it's more than a simple lease getting signed.

Leasing got cheaper, but site acquisition is still costly. Still a problem!

Utilities: Power and Backhaul/Fronthaul

Utilities are expensive to add to anything. There needs to be power, this needs to be trenched or run overhead, and it costs a pretty penny to add a meter, do the permitting, and hire an electrician to wire the pole. Oh, and it needs to be inspected by the city in which, you guessed it, the city gets another fee.

Then there is backhaul, or fronthaul, either way; you need something to get back to the core. You heard me; this was a huge oversight when looking at small cells. You see, when I was working for an OEM we thought we solved this issue using wireless and cable, but the carriers wanted fiber at every site, or at least at that time they did. Fiber is expensive to run to every site, and no matter what the carriers did they could not get the price of fiber to fit the model.

While wireless is excellent, pay once, and it's ready with only power and rent to follow, but the carriers had problems getting connections on those low poles with reasonable bandwidth. The newer fiber systems are allowing more throughput which may resolve the issues at some sites. It's not quite there, but almost.

Then the cable connections, at that time, had such a delay and issues running it through the cable companies core before connecting to the carrier's core. I don't know the specifics, but the carriers didn't like it. Cable companies like to run all the data through their core which causes problems for some carriers. They tend to like dark fiber to a dedicated location. So, the cable option never took off.

What's funny is that fiber is way cheaper and more available than it ever was, but it costs a fortune to run it to a pole especially if you have any trenching to do. Even overhead. The pole rentals add up and the right of ways. Did I mention that you must pull permits for the fiber? That's right, the cities and municipalities are once again involved and collecting money for this service.

I didn't fiber wasn't a great solution, it is, but it costs money to get it to any pole. It ain't free! Not by a long shot. Did I mention that once it's in it goes from a CapEx to an OpEx? Monthly reoccurring costs add up. Even if the carrier owns it, there are attachment fees and rental and maintenance.

Power is expensive! Fiber is expensive! Huge problem! Epic fail for small cells!

What is a Mini Macro?

What is a mini macro? Well, let's look at it as a single sector cell site that you would mount on a pole or up on a rooftop. It would be a standalone site. Softbank used this in Japan. It looks like they will have Sprint deploy them in the USA. Why? It's cheaper than a full-blown cell site! This is a budgetary decision where they are looking for a way to get the biggest bang for their buck.

This mini macro you would have everything you would at a cell site. You would have backhaul, a router, BBU, RRH, antenna, HybriFlex and RF cables. The difference would be that you would just have one compact BBU and one RRH and the backhaul. You need the backhaul to tie the eNodeB into a core.

Let's break it down, on a small pole, monopole, utility pole, or on a rooftop; you would have a very small BTS with the router and optional battery all inclusive. The backhaul could be anything, copper, fiber, or wireless. It would need to be in a small form factor. Well, that sounds like a small cell, doesn't it? The difference is the power will be above 5 watts, probably around 20 watts. It would be just one sector, possibly one antenna. It all depends on the antenna and if you need combiners or filters. These are details that need to be worked out by the OEM and engineering.

Imagine if you will, it will be a small cabinet with or without batteries. It will cover more than a small cell. You will want to have a little more height to get the biggest bang for your buck. The key is to maximize signal for the least amount of money possible.

If the deployment is managed right, I would think the mini macro would be put together as a unit and then deployed as kits, so it should be very close to plug and play. The power would need to be connected, so you still need AC or DC power. The RRH and antenna would need to be connected so you still need the fronthaul cable, HybriFlex or something with the connection to the antenna which could be a direct connect or RF jumper. You need backhaul, let's say wireless, so it must be connected and aligned.

In my world, the BTS would power up, the backhaul would be connected, and then the station would come up and be integrated. The commissioning should be just like a small cell, so it should be ready for plug and play or pre-commissioned. Integration would be done remotely while the installers are still on site. Then the installers could test the BTS and verify the unit is working by testing it with a smartphone device, just like they do for small cells now. The optimization might be an ongoing process, but this also should be monitored remotely and adjusted automatically.

OK, I know this was a high-level explanation, but I think you get the picture, right? What I didn't tell you is that this is the Sprint plan for densification is just like what Softbank did over in Japan. They had great success over there in getting these deployed and covering a densely populated region. This is probably the plan over here now that Softbank is taking over Sprint. I just hope they remember KISS, Keep It Simple Stupid! I think that looking at the Network Vision deployment you realize how complicated it became. If the mini-macro can be simple to install and simple to turn up and simple to test, you have a winning combination. Network Vision was anything but simple for most of the deployment teams that I talked too. It was a huge learning curve for many of them. When going to the lowest bidder, it's hard to get experienced crews repeatedly that can handle something like that.

Figure 6 Non- Pen Mount Picture

Small Cell and CRAN Deployment Report

Figure 7 Outdoor Cabinet Picture

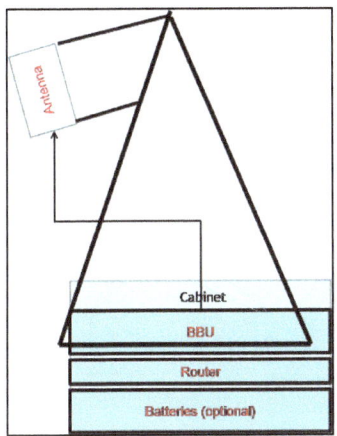

Figure 8 Mini Macro Block Diagram

Why Install Mini Macro Cells?

Because you want more power and more penetration in a specific area! If you need just a bit more power to get the building penetration that you didn't have with 5 watts, then you go with something that has 20 watts. The key here is to get the biggest bang for your buck for coverage and penetration. You probably want more height to cover a bigger area, and the goal is to get more building penetration from the outside in. This is what you will do to avoid putting more equipment in the buildings if possible.

The rise of CRAN and C-RAN/cRAN

Many small cell deployments are just the installation of a remote radio head. Technically, that is not a small cell but a CRAN system. CRAN is a system where a centrally located BBU is controlling many remote radio heads all over a town or a venue. You have the macro benefits of the installation of a small cell radio head, in most cases.

What is CRAN and cRAN/C-RAN?

Here we get into a bit of discussion. You see, CRAN is generally Concentrated Radio Access Network. However, now we are looking at Cloud RAN, which could be C-RAN or cRAN. All the same. We don't have enough letters to break them apart. So many people look at CRAN as a remote RAN, could be a remote radio head or a remote tower site.

CRAN is where the radio head and antenna are remotely placed away from the BBUs. When I say remote, it means that they could be hundreds of kilometers apart, not at the base of a tower. The CRAN system is connected by fiber, muxes, and routers. Generally, dark fiber is to be lit. The BBU is controlling the radio heads altogether, and the link is its lifeline.

cRAN or C-RAN, (cloud RAN), is the same as above but the cloud will be running part of the BBU functions. This means that the cloud could be running some BBU functions to control the radio head and the BBU could offload some processing power.

Why the distinction? Because the cloud RAN is not working quite the way the OEMs had hoped. The BBU is being asked to do more and more specific functions that connect people's data and calls to a specific area. This is no easy task.

CRAN and cRAN/C-RAN are similar in the fact that the radio head(s) stand alone.

Small Cell and CRAN Deployment Report

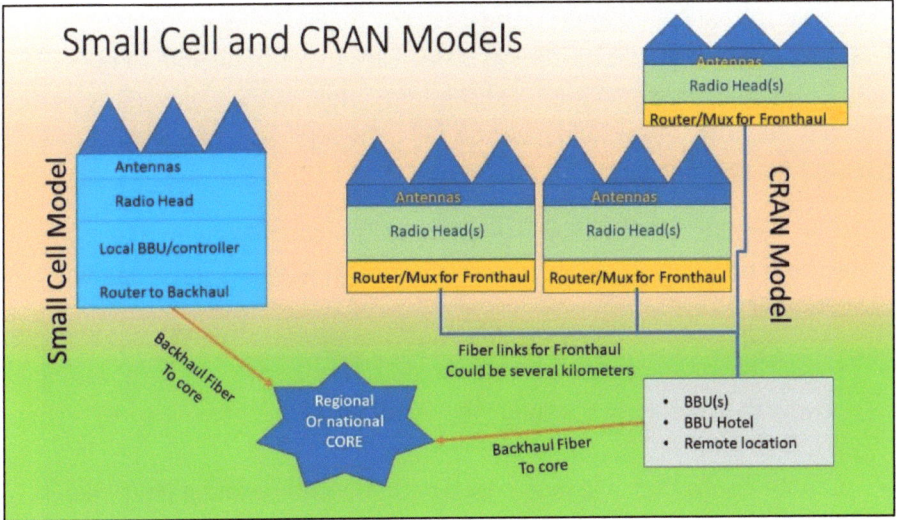

Figure 9 Small Cell and CRAN Models

Learn CRAN at:

- https://wade4wireless.com/2018/02/19/who-has-enough-spectrum-for-5g/
- https://wade4wireless.com/2015/07/27/small-cell-fronthaul-and-odas/
- https://wade4wireless.com/2015/11/12/wireless-deployment-handbook-for-lte-small-cells-and-das/
- http://www.senzafiliconsulting.com/Blog/tabid/64/articleType/ArticleView/articleId/229/C-RAN-Airvana-Taking-C-RAN-indoors-to-optimize-small-cell-performance.aspx
- https://dailywirelessnews.com/wp-content/uploads/2017/12/Mavenir-Whitepaper-5G-2017.pdf
- https://www.rcrwireless.com/20151222/featured/what-is-c-ran-tag4
- https://www.rcrwireless.com/20160803/analyst-angle/analyst-angle-cran-stand-anyhow-tag9
- https://www.commscope.com/Solutions/What-Are-C-RAN-Small-Cells/
- https://www.sdxcentral.com/5g/definitions/cran/

What is the difference between a small cell and CRAN?

I thought you would never ask! The difference is not only is the small cell a stand-alone unit that just needs backhaul to a controller or core, but it generally is lower power. It is also limited in the throughput it can handle.

CRAN is usually a larger radio head with full macro functionality and loading because it is connected to a macro BBU. It can do everything that a macro sector can do. The only limitation is the BBU, not the radio head. If you need 3 carriers, the BBU can handle it whereas a small cell may not be able to do 2 or 3. A macro BBU can handle many users at the same time, but a true small cell is very limited. A CRAN setup could give you full blow services and coverage in a limited area for similar site costs.

To sum it up, the CRAN radio head has full macro functionality with a seriously dedicated fiber backhaul whilst a small cell is lower power and lower loading stand-alone unit.

- Small cell, lower power, all-inclusive unit, less loading, and processing power than a macro. Backhaul just needs to connect to the core; timing is not so critical.
- CRAN – controlled by a macro BBU with full loading and higher power units, full macro capability. Backhaul is critical, needs to have low latency and connected to the BBU, then the BBU will connect to the core.

What are the advantages and disadvantages of small cells?

The small cell can be used for 2 reasons; they can be put in for coverage or for offloading a macro site. Keep both in mind when reading the next part.

- Advantages are:
    - The device is smaller, cheaper, versatile.
    - Can offload traffic from the macro?
    - Cheaper to install, deploy, maintain.
    - Backhaul is critical but more versatile. Can be put many miles from the core.
- Disadvantages are:
    - Payback is low.
    - Limited loading capability.
    - Generally low power, (could be advantage or disadvantage).
    - Limited coverage.

What are the advantages and disadvantages of CRAN?

The CRAN is generally a full function macro sector, (or sectors) in the system. Remember that this is a macro system that has been split to serve multiple areas at the same time.

- Advantages are:

- - Full macro capability.
  - Loading is that of a macro.
  - Full power is possible at a site.
  - Complete macro functionality at the radio head.
- Disadvantages are:
  - The limited distance between the radio head and the BBU, several kilometers depending on the OEM.
  - Fronthaul to the BBU is critical, low latency, dark fiber, limited distance, higher cost.
  - Very expensive to deploy a new system.
  - Planning matters because you're connecting to a BBU nearby.

How do I choose which to deploy?

This is the big question. All the stuff above will help you weigh in on this decision, but here is the real head-scratcher. When do I deploy a CRAN system and when do I deploy a small cell? Let's look at the scenarios below and make some decisions.

Scenarios:

Looking at each scenario, let's see how each one plays out.

- Will loading be an issue? Is this a city with a lot of traffic that will have heavy data and voice traffic?
  - If yes, then CRAN is a better option since it can take loading of subscribers and data from a macro site.
  - If no or the answer is sometimes, then small cells may be a more affordable option with a better payback.
- Is coverage the issue outdoors?
  - If yes and you need a higher power and have a higher site, (need to be sure the public is safe from an RF radiation), then a CRAN would be a great option.
  - If no and you are covering a very specific are like a town square or indoors, then a small cell makes a lot of sense.
- Are you going to deploy across the entire city or are?
  - If yes and you need to deploy for loading and coverage, then the CRAN is the ideal option.
  - If no and you are only covering a specific area, then small cells would work.
- What if I need indoor coverage?

- - If this is a convention center or a stadium, then CRAN makes a lot of sense because of the massive loading.
    - If it's a business or office building or train or bus station that is not always busy, then small cells make the most sense.
  - Does spectrum reuse an issue?
    - If so, the CRAN makes a lot of sense since it is controlled by a common BBU.

What about the cost and payback?

Here is what you need to understand, the cost differences between each system and the payback. This will help you understand why the scenario makes a difference.

CRAN systems and costs:

The CRAN system is more than just one macro system in an area. It consists of several BBUs and a large fiber network. The system must be thought out for an area or a stadium. It's not something you just throw in with little thought. The system should be planned for the job it must do.

- The system will have
  - **BBU site** will be in one spot in a city or stadium - A set of BBUs, called a BBU pool or a BBU hotel all in one area. It's more than the BBUs; it is a power plant, backup power which is generally a generator and batteries, a complete set of fiber muxes that go out to the radio heads as well as a larger, 10Gbps or higher, backhaul. All of this in one large closet that needs to have the temperature maintained. It will also have servers and other equipment in there to support the BBUs. You could have 1 BBU to 10s of BBUs all in one space connecting to many radio heads across a region or stadium.
  - **Remote radio head** – could have the radios heads at a tower with 3 sectors, this is not common yet, but the industry wants to get there to get equipment off the ground. The more common scenario is to have a single or double radio head on a pole in the city that gives it macro like coverage for loading and coverage. This goes beyond the densification to serve as a smaller macro site with full macro capability. You still need power, possibly batteries, and routers at the radio head site.

You will have antennas to connect to the radio heads. Fiber to the router, then fiber to the radio head, then coax to the antenna, unless it is an active antenna like what massive MIMO will use.

- **Backhaul** – this runs from the BBU site to the core and should be 10Gbps or more.
- **Fronthaul** – this is generally run from the BBU site to each individual radio head site. It will take a series of muxes to send out the signals across fiber to remote radio heads. The latency between the BBU and remote radio head is critical; there is a distance limitation due to the timing of the signals from the BBU to the radio head. While this is being improved by the OEMs, it is still a limiting factor.
- **Fiber** – Fiber is needed to make all the connections. It is critical.
- **Fiber Mux** – The signals need to have some type of multiplexer system to manage the signals. Remember at the BBU side there will be lines feeding many radio heads, so there must be some sophisticated way to manage all the outgoing and incoming lines.

As you can guess, this is very expensive to install. The payback could look like this; one CRAN system could replace tower sites across the city. It would also allow heavy loading at each location. This is an ideal setup for parts of major cities where there are no towers or accessible rooftop. It is also an ideal solution for areas that have heavy loading, like a large stadium with over 30,000 people. In a stadium, you may be able to connect the fiber directly from the BBU to the remote radio head but check with your OEM.

Small Cell systems and costs:

As you have probably guessed, small cells are much cheaper and easier to deploy. They generally are easier to plan to deploy.

- The system will have:
    - **A small cell** - with an antenna, could be integrated or could be a separate antenna.
    - **Backhaul** – this could be dark fiber lit for this purpose, could be shared backhaul or could be a VPN back to the core or server, depending on the application.

That's it, easy and cheap considering. I can't do much about the site acquisition part, that always cost a lot of money. However, the small cells are generally very cost effective. They cost less to install, and the engineering is generally very reasonable.

What's the cost difference?

Here is how to look at this when concerning the cost difference. Therefore, the payback is critical. It's the difference between putting in a macro and a small cell site. Keep in mind that these are all estimates, not hard costs. Prices differ depending on OEM, region, design, planning, loading, coverage, rent, and so on.

Please note, these are all very high-level estimates!

- Small cells
    - Equipment, generally less than $15K per site
    - Physical installation, generally less than $5K per site
    - Backhaul, generally $5K to $15K to install and could be $300 to $3K per month.
    - Coverage, less than .1 miles or a room in a building.
    - Rent, generally $30 to $3K a month. It really does vary that much!
- CRAN
    - BBU site equipment - $30 to $100K.
    - BBU Site Installation - $3K to $50K.
    - BBU site utility power - $3K to $13K.
    - BBU site routers, muxes, and backhaul/fronthaul equipment generally $3K to $50K, maybe more depending on load.
    - BBU site rent is generally $2K to $$10K each month.
    - Remote radio head equipment $10K to $30K with antenna and hardware
    - Remote Radio routers generally $1K to $5K
    - Remote radio head installation generally $1K to $5K unless it's a tower
    - Remote Radio Head rent could be $500 to $3K each month unless it's a remote tower, then it's $1K to $3K each month.
    - Backhaul and fronthaul connection generally $1K to $50K each month depending on how many radio heads, substantial backhaul, and all the dark fiber needed to be lit.

Small Cell and CRAN Deployment Report

There you go! It's not as tight as you would like it, but you get the idea. CRAN is expensive, but the payback is there for the situation that you need it.

However, as attractive as the small cell looks, it's payback is not there. The small cells serve a purpose as an adder, but not as the primary system. That's why the small cell is an excellent solution for its purpose which could be to offload the macro site or act as a fill site. It's a great indoor solution.

If you don't know it, the CRAN systems are a secret weapon of the carriers here in the USA. You see, the CRAN offers a lot more diversity and loading, all the features of a macro site and the payback is far higher. It is the system that allows the radio heads to handle thousands of users at a time versus a small cell which handles hundreds at a time.

Would you Deploy Small Cells or CRAN?

This is an ongoing question that many people have been asking, with the shape of the industry rounding out to be 5G and massive MIMO, "How each application works and why?'. If you were to make an investment with your limited budget, what would it be, small cells or CRAN? Of course, it's not black and white; you have an application for each one. That's what I will discuss below.

Would you deploy DAS or Small Cells?

Both are a great solution, and each carrier will have their own strategy. I think that some carriers may rely on macro for the deployment but let's face it, they need to get the signal to the people and chances are they already tested both small cell and DAS options. It all comes down to strategy and budgets.

Normally, contractors won't have a say; the carriers already know how they want to deploy. The carriers have a strategy. It differs from carrier to carrier. It differs from venue to venue. They may want DAS inside and at a stadium but out in the street they may want small cells. Or it may depend on the customers they are trying to hit. They plan around the expense and the customer and the need for that venue. I know that many indoor carriers just send out a femtocell to cover that space because it is easy and cheap. T-Mobile will do this for many people, and the femto will have Wi-Fi in it along with the carrier.

The value contractors may add to recommend how they should deploy in a venue. If they are asking, then make sure you evaluate what the costs are versus the expected coverage. If they are looking for offloading, then make sure you not only deploy the solution but make the right call for backhaul.

The difference here will be cost and payback. I don't think the carriers will put any money into small DAS systems anymore. They will go the small cell or CRAN route. They know that DAS cost a lot of money to install and maintain. It makes the most sense to put it into a larger venue but not small venues.

That is why the carriers publicly said that they wouldn't support smaller DAS systems. They are stuck in the stadiums doing what they should. The chances are good that they will only deploy digital in the future. Why install a small DAS when a small cell works just fine?

Carrier DAS systems continue to be deployed and upgraded in the USA, but only when they make sense. It is a huge cost to install and maintain. Upgrades can be painful. DAS is needed in larger venues, but it's looking more and more like CRAN. They use digital DAS more and more. The antenna distribution will become more and more critical. The growth of 5G will also make it easier for CRAN to work at this venue because it should allow more loading on one antenna. The other kicker for DAS will be massive MIMO, where there is an active antenna. It's hard to share a system when each carrier wants their own active antenna at the venue.

That means that a common carrier small cell will be hard and hard to find. We'll need to rely on LTE-U or CBRS to have common carrier devices. While this sounds great, I would expect that the performance will be diminished.

Do you need to choose?

I will let you in on a secret. When small cells came out, the DAS vendors were worried and saw them as a threat. That's right; the DAS vendors saw small cells replacing all their DAS systems. They did not look at them as a Wi-Fi model, instead of as a replacement of DAS. Nothing could be farther from the truth.

The reality is that DAS systems found out that small cells were their friends. In fact, they supported DAS systems in several ways. They were cheaper to deploy and had multiple functions.

One, they could be used as the head end. While a small cell doesn't have the capacity that a macro site does, it was easy to install at the head end, and it would feed an RF head end. Unfortunately, these models are quickly fading away.

Two, they are deployed in place of radio heads. They are easy to deploy, and they only need a fiber connection or CAT5, depending on the model, to connect the small cell to a router somewhere. It still looks like a DAS system, but it's cheaper and easier to deploy.

First off, the DAS I am talking about is for the carriers, not public safety. Public safety DAS is still going strong to have the bands working properly throughout the building for fireman and police. Those are still mostly analog or passive DAS systems.

The carriers DAS systems have evolved rapidly over the past 5 years. They are all digital systems. The DAS systems had to evolve to talk to digital radio heads.

They should start using CAT5 then fiber to connect to the remote radio heads. In fact, they basically connect to the OEM's radio heads to distribute the signal across any given venue. They may just connect to small cells or radio heads, it all depends on the loading. System loading will determine the design. For instance, if a super bowl is coming up, you can expect heavy loading, but if it's a business center, then it could be steady loading for business hours only.

Will DAS and Small Cells Work Together?

Of course, they love each other in this Heterogeneous Wireless World we work in! Let's start by clarifying that DAS and small cells work very well together. You see, when I talk about the small cell, I really mean an all-inclusive unit. So, the small cell would stand alone without any connection except the backhaul to the core and a connection to the antenna. You still need power, and the antenna may be external, but you get the idea. CRAN would be part of a DAS system meaning it would have a separate BBU to connect to the backhaul and to the radio heads which would attach to the antennas. Other DAS systems could be fed by a BBU or a small cell or a full eNodeB then distribute the RF signal by feeding antennas with RF cables. CRAN makes the most sense to me in the distribution case.

The small cell could feed a DAS system with lower power, thereby eliminating the attenuators that a full-blown eNodeB would need. Why would a normal DAS system need attenuators? Well, they tried to set power lower on some eNodeBs, and that does work. The issue has been that many eNodeBs will come up at full power after being reset or rebooted, and they may transmit on full power before making any adjustment. This caused the distribution systems to blow out taking down the DAS and incurring very expensive repairs. I mean a huge expense in parts and labor. DAS engineers learned to set it at full power and attenuate it down. If you worked with these systems, then you can feel the heat, and you hope the cooling system does not fail! So, the Small Cells will help eliminate the cooling problem. Lower power, fewer attenuators mean lower costs and lower cooling costs. One more thing, fans fail often and need to be cared for or replaced regularly. Removing the fans will remove a point of failure.

In the Het Net, you have the small cell and the DAS system when the small cell can feed the DAS system saving money on the installation and rent because you need less space. This is what many DAS vendors are looking forward to, a way to save on OpEx, (operating expenses, monthly costs), and CapEx, (capital expenses, installation, and hardware costs). However, some DAS vendors like TE

Connectivity, (acquired by CommScope), made a common interface to their radio units using CPRI. The setup they have is that an OEM, (like Alcatel-Lucent who signed a deal with TE), can provide a BBU that will feed the DAS system that TE built. Then the DAS system can distribute the CPRI to the remote radios to feed a building or venue. Sound like CRAN? It does to me, and that is the basis of most modern DAS deployments. This is an active DAS system.

Depending on the venue the DAS system may need to transmit more than one band, they could include several carriers and maybe even Wi-Fi. A Small Cell is usually dedicated to one carrier and may or may not have more than one band. Usually, it is one band with multiple channels within the carrier's arsenal. The multiband small cell is still being worked on, but in most cases, they deploy multiple small cells. The small cell is normally dedicated to one technology, like 3G or 4G, but they could include both. Remember that the small cell is usually a very small cell site, so it has limitations. Whereas DAS will have the flexibility for the customer to use any carrier and get Wi-Fi access all at the same time, assuming it is set up that way. Small Cells probably will have Wi-Fi co-located in them or near them, but really, let's take a small cell for what it is, a small cell site, and Wi-Fi for what it is, a hot spot. I was reading Martha DeGrasse's RCR article about Het Net and DAS, found at http://www.rcrwireless.com/20140929/hetnet-news/hetnet-news-lte-drives-das-upgrades-tag4, and it reminded me how much work they have for big venues, like the Super Bowl. You can see the 2014 Super Bowl results in Fierce Wireless at http://www.fiercewireless.com/special-reports/super-bowl-xlviii-how-did-tier-1-wireless-carriers-networks-hold. Fierce Wireless wrote about the deployment issues http://www.fiercewireless.com/story/madden-das-or-small-cells-case-study/2014-04-28 if you want to see the limitations out there.

Again, next time you talk to a DAS deployment team, shake their hand and say, "Good job!" so they know they are appreciated. It is no easy task to plan and deploy and test. Upgrading is even more of an issue!

DAS and CRAN

Most digital DAS systems for carriers are using CRAN. They generally connect to a BBU for a carrier, located in that building, and control a radio head near an antenna somewhere else. It's that simple. They use fiber for a fronthaul to connect to many radio heads from a BBU hotel.

Small Cell and CRAN Deployment Report

This is the way it's been done for 4G and beyond. In 4G and 5G there is not passive system or amplifiers that work. Generally, it must go through the radio head and direct to an antenna. Carriers are no happy with combiners or slitter as much as they use to. To get the performance that is expected, they will need to have a dedicated radio head that feeds an antenna or combiner direct. They don't want to rely on too many passive devices to degrade performance.

Digital DAS and CRAN are very close to the same thing. It's just that the DAS tries to support more than one carriers. In fact, when you hear of oDAS, chances are it's a CRAN model. If you hear of a carrier doing DAS in a large venue, chances are its CRAN.

What G makes sense for small cells?

Something that should be covered is the technology used in the deployments, whether CRAN or small cell. Could it be 2G, 3G, 4G, or 5G or a combination of all of them?

2g? Are you kidding me?

I am going to speak mostly about the USA here, because 2G, while it's out there, will be decommissioned soon. I don't see anyone putting money into 2G. As a matter of fact, I see the carriers planning to take the 2G spectrum and convert it into IOT spectrum so that they can responsibly call it 5G. Now they have a use for that narrow bandwidth spectrum that serves little purpose in today's heavy data-centric world and makes money off it in IOT. They won't have to give it back to the FCC to pay for it again. Instead, they could use it for the IOT deployments. They need new equipment, make it LTE, and make it very low latency.

3G was not hot either.

3G was a better option for small cells because it did voice and data. It was CDMA or GSM, but most companies still used repeaters to pass the signal and improve coverage. Loading was not as much of an issue.

4G and LTE make Sense.

Now, 4G is all LTE and data-centric. It makes the most sense to have an all-digital small cell because it offers you the loading solution. With LTE the expectation is to offer heavy data. The small cell helps with the offloading and coverage. With the growth of smartphones, it changed the role of small cells. They are not being asked to be a new hot spot and for offloading the macro site.

And now, 5G!

With 5G rolling out, it makes the small cells even more important in the ecosystem. At least a small cell or CRAN because of the broadband requirements and the "coverage everywhere" requirement. This will begin to shape the small cell industry to install 5G like we used to do Wi-Fi. It's going to be needed everywhere just to keep up with the need.

Not only that, but IOT is a huge part of 5G. IOT is a way to connect devices to the network, which small cells may be a very cost-effective way to connect many devices all at the same time. You have to be where the devices are. I am not sure about 5G, but there are LTE formats that use little battery. That's key if connecting remote devices. You should have something out where the devices are, so they can talk and hear. They have to connect somehow.

The small cell will become more important in 5G for the coverage and for the spectrum. As higher bands, like mmwave, are used, small cells can produce suitable coverage for the users. It can become the connection for devices where there is a concentration of people or devices.

You still need backhaul, so if it's for broadband, they may opt to go with a CRAN solution since it may have the capability to handle more throughput than a standard small cell.

Again, the solution will determine what will be installed. IOT may be narrowband and low bandwidth, maybe even lower spectrum as it travels farther. Massive broadband solutions may be mmwave solutions that can deliver the bandwidth needed. Either way, a remote radio head near the people will be needed.

Small Cell Opportunities

Deployment is relatively easy for the installation and integration teams. They have become easy and straightforward. At the site, the backhaul is still an issue, and site acquisition is still costly. Not only that, but the planning has to be taken very seriously now. At one-time, design teams thought they could deploy easy and quick, but it seems that asset owners have different ideas. Things don't roll out overnight, and the carriers can't blame the OEMs or installation teams for high costs. No, they can blame those pesky backhaul providers. Carriers make a lot of money, but they do all that they can to cut expenses everywhere.

Installation teams expected more work than what they got, but it never ramped up. It's ironic, but the companies that would get the least amount of money were hoping for the surge of work. Don't get me wrong, there are a lot of small cells out there, thousands. There are many CRAN installations where someone like Verizon or AT&T would install the radio heads that go back to a BBU hotel somewhere in town. Also, Sprint and T-Mobile found a way to use Mini Macros as fill sites. They found that it's cheaper to deploy and can fill many holes that a macro was too expensive to fill. While not the best solution, it works, and it's cost-effective. It's another tool in the arsenal of deployment solutions.

Small Cell and CRAN Deployment Report

So instead of 100s of thousands, there were thousands, and they were scattered. Never a mass deployment across any city except maybe large cities like New York City, LA, Miami, Chicago, Boston, San Francisco, or Philadelphia. Still, not the masses we hoped for.

The opportunities are there. All we should do is look at what Crown Castle (CCI) has done. They have locked in the sites and added fiber to the site as a value-add. While states limiting rent on city poles model may have hurt the leasing model that CCI, they will make it up with the fiber and utility costs. They got smart with the package concept and saw a future there. Those people at Crown Castle have vision!

The new spectrum using mmwave and cmwave may be more than a small cell but the backhaul solutions needed to deliver backhaul form a small cell. New 5G technologies that the carriers are toting for broadband may resolve many backhaul and fronthaul issues that plagued them in the past. Extending fiber with wireless connections that can do 1Gbps to 10Gbps reliably is key to success. It will make the backhaul economical, especially if they are multipoint. I believe the line of sight will still be an issue but if it works, use it. So, when thinking of small cells, they will now be used for more than connecting the end user. They could be used to replace a cable modem in a home or as backhaul to another small cell or even for a small business' internet connection.

Using UE backhaul like Sprint has done successfully is going to happen more and more. With the throughput increasing then you have a way to hop from cell to cell and bypass fiber altogether. Sprint is trying this model, and as they improve their 2.5GHz roll out across the USA, it may become a viable solution for expanding coverage. Finally, Sprint did something right with that spectrum! Just kidding, but it seems like they took their time utilizing it for much of anything on a grand scale.

Looking at the macro as a feed for small cell backhaul makes a lot of sense. The macro has a large backhaul connection and more equipment than any small cell site, so using it as a hub creates new possibilities for extended midhaul and fronthaul.

Indoors we need small cells more than ever. I know most carriers try to go from the outside in, good luck with that. You need something indoors, and the DAS systems are costly in smaller venues. We all heard the carriers say they didn't want to pay for it, but they will need indoor small cells. It must happen. We need indoor small cells to cover buildings. There is a point where a massive DAS system will be needed, but in most cases, splattering small cells across an area will do the trick.

Overview by Market:

Here we have the growth of the market. It's easy to look at the market overall and take a guess. The reality is that we need to break down small cells in specific markets. When I mention small cell markets, I look at a breakdown of enterprise, indoor, and outdoor. This way we can look not only what vendors will be deploying but who will be purchasing the small cell. I purposely skipped over the home femtocell market because it's more like a UE device. I know what you're thinking, it's not, sort of, but it is. It is for home use, and self-install and the only deployment happening with a femtocell is going to be done by the end customer possibly with phone support. That's how the carriers can take credit for so many small cells being deployed. They often look at the total number of small cells instead of breaking it down. So, remember that when looking at this I am not looking at the femtocells.

Enterprise:

Most enterprise small cells are looking to add coverage in their offices. The thing is that the carriers don't want to supplement the enterprise installations or equipment any more than they should. The enterprise is going to have to pay for it themselves. Many will go with the carrier solutions if the price is right. Many enterprise businesses are going to settle for the Wi-Fi that they can install cost-effectively and easily. Let's face it; Wi-Fi is super easy to install.

However, if the FCC ever gets off their ass and finalizes the CBRS rule, chances are good they will go with the 3.5GHz lightly licensed band if the devices will have it available. Why? Security and ease of deployment. It's one thing that you can do licensed without a major roll out. It's quick, easy, and efficient. They can use it with Wi-Fi.

What does this mean for vendors? It means that SpiderCloud will remain on top for enterprise small cells. They have worked hard to get to that position. It seems like they want the enterprise market more than the big boys, like Nokia and Ericsson. The big boys didn't make a true effort in the enterprise, mainly because the carriers didn't make any effort. In fact, the carriers don't push it because they don't want to pay for the installation.

Small Cell and CRAN Deployment Report

For the enterprise, private LTE systems look very attractive. It will remove some of the limitations that Wi-Fi has if the company is willing to invest the time and money to turn it up. The CBRS will make it easy from a spectrum perspective. That is going to make it easier for the company to roll out new private LTE systems. It makes sense because of the characteristics of LTE are better than Wi-Fi for connecting. It's also going to be a clean handoff for the end user's device. The company could have its own network independent of the carrier. If the carrier doesn't pay to install small cells on your business campus, then make your system a system that they could roam onto. Maybe they would have to pay the company a roaming charge. Then the carrier would have to make a deal to roam onto each private network. They could use LTE-U or CBRS spectrum. It makes sense.

Indoor:

Let's keep this separate from the enterprise. Granted, the enterprise is almost always indoors unless they are trying to cover a campus. Then it would be a mix but, in most cases, they cover from the inside out, unlike the carriers which start outside then work their way in. Let's move on.

Indoor is going to continue to grow. It will be a compliment to DAS, even though DAS is still growing in the USA, it is more expansion than new installations. Although, DAS vendors won't say that. They would say it's growing, but if you ask, most are busy with growth to existing locations, not new locations unless they are already too large to deploy small cells cost effectively. This is still the issue.

Let's face it; indoor small cells will continue to grow for public venues. They are cost effective but the venue owner, like many airports I know, do not want to see 6 small cells lined up where the public can see them. That's been an issue, yet, I am not sure how to get around it. They will see the small cell, the antenna, or something unless you have stealth. I think that a fake fiberglass wall is the best thing that could be done. Make it look nice with a drawing or painting on it. Then you can put whatever you want behind it. Unfortunately, the installers may be on the hook for that. It will cost more, but the benefit should outweigh the cost.

So, when doing an installation like an airport or a large building like a business plaza make sure that the aesthetics are looking great. Make an effort to have it look good to the consumer, the person walking around in the lobby.

Small Cell and CRAN Deployment Report

Why a small cell over a DAS system? When the loading will not be an issue. A small cell can handle a few users, let's say up to 30, or maybe it's a bandwidth and loading issue, meaning that the users are heavy data users. If they are heavy data users, like what you may see at a convention center, then DAS makes sense because a macro site can handle an enormous amount of data. Then the DAS system with a macro BBU makes a lot of sense. That's why the NFL stadiums have DAS systems that are growing. Here is where smaller CRAN systems come in.

The CRAN system is really oDAS, with a central BBU hotel that is completely controlled by the carrier's BBU and fiber connections. These are commonly used indoors and out. However, it is all the carrier, meaning the carrier will usually control everything. When looking at a venue, indoors or a stadium, the venue owner generally wants 1 manager of the wireless system. They generally partner with a DAS manager that will connect the BBU to the radio heads or distribute the antennas across the stadium. More stadiums rely on the carriers to maintain their own systems and they have a company manage the DAS. It's pay to play for the carriers.

If a regular building owner wants coverage in the building, then they will need to add something. Currently, they have Wi-Fi, because it's cheap and easy and usually you can find a WISP that will maintain it. Soon, it could be a CBRS or LTE-U hotspot maintained by a WISP. One thing I learned is that the building owner will not install anything unless they have too, then it must be very cost-effective, (cheap).

I remember talking to the team that managed the wireless for the Las Vegas Convention center. I believe it's managed by Cox; information found here https://www.nbaa.org/events/bace/2015/general-info/NBAA2015-LVCC-Cox-WiFi.pdf if you're interested. They oversee the Wi-Fi and the management of it, but I believe they also manage the carriers who have DAS and radio heads throughout the venue. You see, for the convention center to deal with all the carriers and the Wi-Fi vendors is a headache for them. After all, their business is booking the convention center and maintaining it. Wireless is one part of it. They have a team to manage the wireless. Under the wireless umbrella are Wi-Fi and Cellular. The cellular, in this case, is all the carriers who have coverage in the building. It really is impressive the way they have the systems installed to cover the entire venue.

It's more than Wi-Fi, it is also the tracking of users, the counting of people in the venue and walking by. The managing of all the heavy data usage. The Cox team does an amazing job managing everything, physical and data. It's no easy task.

If you think that they get a day off, guess again, they usually work holidays to make sure that the upgrades and changes are made. No one said wireless gets a day off. Holidays are when the teams do upgrades, additions, and changes. That is how these venues continue to grow and improve.

So, the indoor venues are always trying to add where they can. I am making this point because what you may think is a small cell could be a radio head connected to a BBU hotel somewhere else.

When thinking of indoor small cells, we often shortchange ourselves. What we think it that the small cell is the only thing to be installed. If you're an investor looking at the hardware and the deployment, it looks reasonable. But if you own the venue, you need to think beyond small cells, Wi-Fi, and coverage. Think about the applications. What can you offer beyond the small cell, but I digress, that is another chapter in the book.

Outdoor:

Outdoor small cells are not taking off unless you count the CRAN market as a small cell, then it has a chance of taking off. This is different from a real small cell because you're putting out a radio head, not a full blown small cell. There is a difference, but to the installer, they all seem the same. It depends on your perspective.

Back to CRAN, this is what is generally associated with outdoor DAS, oDAS. The reason I bring this up in the outdoor small cells section is that many installers look at the radio head installation as a small cell installation. So, it gets lumped into this section even though it's not a small cell, but all the things that people measure a small cell on are the radio head installation.

The difference between the small cell and the CRAN can be described like this; a small cell is an all-inclusive unit that connects to backhaul and is small cell site. The CRAN is 2 separate units, the BBU hotel at a remote location and the radio head mounted on a remote structure connected to fronthaul.

I often see confusion when people, investors, engineers, governments, and installers talk about small cells. I am just going to say that installing the radio head is the same as a small cell to the structure would be treated as a small cell.

In fact, the only people who care about the difference are the engineers at the carrier and possibly the fiber provider. As for the remote end, it all looks the same. You have the heavy unit, the radio head or small cell, the antenna, and the backhaul or fronthaul connection which is the router and the fiber connection. It all looks the same from that perspective.

Why would the engineers care? Here's why they need to worry about how to set up the unit. If it's a small cell, then it needs to be set up like a cell site, but if it's a CRAN radio head, then the BBU needs to know where the radio head is, how far away it is, and what it can hear. While the small cell sounds much easier to deploy, and it is, the CRAN offers technology that makes it more attractive to the carriers. Remember, this is not a simple Wi-Fi deployment that you install and hope for the best, it is a carrier deployment where dropped calls mean lost customers. The CRAN system offers the control to be done at one central location, the neighbor list by sector is controlled, and the throughput and loading are way better than the average small cell. Looking at it from a techie's point of view, the CRAN system offers so many advantages that could be cost-effective in a city environment where the loading is determined by walking traffic and nearby vehicles. Whereas the small cell may just be enough to cover the walking traffic but if there is a lot of traffic driving by it may not be able to support them as well as the walkers.

Planning for the system is what carriers do so well. They come up with a plan based on what they know from the macro site that is nearby or the complaints they had or some other data that they analyzed to prepare for the install. Some use a copycat technique that is easy, they look at what the other guy did in that area and do the same, no analytics needed, just do what they did because they did that for a reason, right? Easy peasy, lemon squeezy.

The real winners:

The real winner here will be the fiber deployment teams. They will make the most money because even where there is a wireless backhaul, there is fiber somewhere. Even at the site, they use fiber for all the connectivity. Fiber everywhere. Backhaul fiber, fiber jumpers, fiber solutions, it all adds up to success in the fiber industry.

Then come the OEMs, they will sell a lot of product. However, there is a lot of competition. All the OEMs have small cells products, and they are all over the board and spectrum. There's a lot of competition.

Antenna makers should make money here, for both the small cell and the wireless backhaul. It's a viable industry that has limited competition.

Finally, the wireless backhaul provider should do quite well in the small cell boom if they have a small but robust system. Also, a niche market, like a UE relay that works, like what Airspan has done for Sprint, this is a great market to get into. I see mass deployments there.

Why not just put in more macro sites?

This is the question that was asked early on. The macro sites can do more and do a great job. The issue is the cost and how you want to cover. To put macro sites in everywhere doesn't make economic sense when a small cell could cover a limited area, especially for offloading.

Macro sites are a major expense like I said earlier. They don't always cover the way that we hope, but they are the best way to build out a system. The carrier still has to get the biggest band for his buck. So, the issue is, whom are we serving and why? If it's a concentrated area and it's mostly for coverage or offloading, then it makes sense to use a small cell to save on cost.

If it's a large area for coverage, maybe a macro or mini macro. If it's larger area with a dense population, macro with the offloading of small cells.

There are so many options that can be designed for each scenario.

What about the Massive MIMO Macro?

The one thing that will hurt small cells in the urban area, hypothetically, is the massive MIMO macro site. If these multi-user systems work the way that the OEMs say it will, it will increase densification. These systems should offer high throughput to multiple users simultaneously. If the backhaul is good and the beamforming works the way we expect it too, then the offloading in the umbrella of the small cell may not be as critical as we once thought. Small cells are constantly used to offload macro sites because it loads s macro sector down. Massive MIMO should take care of this problem.

Small cells are still needed, but if the beamforming can locate heavy users and concentrate on them without harming other users' coverage or downloads, then the system will allow a macro site to do much more than it could before. It takes away the need for small cells in the massive MIMO sector's coverage.

While this may or may not hurt the small cell plan, it doesn't set it back too much. Users are data-centric, they always want more. The macro site may not have enough fiber or could easily get overloaded with the new 5G requirements. It all piles on to what the carrier is trying to offer.

Small Cell and CRAN Deployment Report

How will 5G that change the small cell model?

LTE was a breakthrough for small cells because it was completely digital. That makes the small cell more and more efficient. It enabled the backhaul to be digital and radio to be digital. This makes it easier to design a small cell. With the sunset of 3G and 2G, it also makes the small cell more and more relevant. Now that the carrier can use a complete digital carrier, things get easy. With 5G there will be one large carrier, not all the smaller carriers that LTE had. It makes the radio capable of doing more with less.

Sure, 5G is new, but it opens new doors for the small cell's capability to pass more data. It improves coverage and it densification.

Extreme Broadband

While we all hoped small cells would be the hero here, it's not likely. This is where CRAN will have to save the day. Most small cells are very limited in what they can pass traffic wise. They may do a good job with one or two customers but get many more, and it must allocate resources accordingly. It's not always an easy thing. Broadband for most carriers will require aggregation of many carriers to get enough spectrum to pass broadband.

However, let's not count small cells out. We are counting on them to assist the macro site with LAA. This is where they will support broadband. They will assist the macro site in the delivery of broadband. For LWA it will be Wi-Fi, but not a great solution. For all LTE systems, it will be LAA of aggregation of multiple carriers at the macro site along with the small cell. So, they will play a part, but I see CRAN playing a larger role since it can do all of that along with the macro functionality.

The CBRS spectrum is where the small cell can really add something. IT can assist the macro and offer some additional coverage in the spectrum that the carriers may not have at the macro site. It's a nice play to have the small cells assist the macro and each other if needed.

Ultra-Reliable Low Latency

Small cells will do a great job here. They can be close to the user and provide a connection direct to the internet or whatever server they are relying on to complete the connection. This is where SDN and NFV will help lower the latency along with a direct internet connection. A small cell could drastically cut down on latency and improve the coverage to the mission-critical equipment. If you put it nearby, it may solve the problems that a macro would have. Macros can get overload because they cover a larger area, but the small cell could be deployed specifically for this purpose. It would have a limited coverage area, so loading may not be an issue. It could even be dedicated to a few devices.

Let's say you had vehicles that needed to be connected for a trip; small cells could be near the highway to keep constant communication open. This way you would not rely on a macro site to stay connected in a dense urban area. A network of well-placed small cells could lower latency and dedicate links for that use.

Massive IOT Connectivity

Here is another place I could see small cells shine, but we must be careful. Most small cells cannot handle the loading that we think will happen. But here, when they say massive IOT, that means there will be a lot of devices, but they will be spread all over an area, not necessarily concentrated in one spot. That would not make a lot of sense.

This is a case where the economics of having small cells spread all over to cover many devices makes a lot of sense. It would be the extended reach that could make it valuable. Then the backhaul for these devices should not need fiber or even more than 1Mbps in many cases. These will be spurts of data, perhaps at given times. Power and backhaul could be the bare minimum. It goes back to the days when utilities used remote control, called SCADA systems, (SCADA for Supervisory Control and Data Acquisition) systems used low power links to connect back to a control center. Same concept here only now a carrier could manage the link and the backhaul. It sounds great hypothetically, but utilities may want full control which will force them to build their own networks, but that is another topic for another time.

The point here is that small cells are a cost-effective way to control a large area with narrow broadband. This is an ideal situation if you don't overload any individual small cell with users. I don't see that happening, but you never know what the future will bring.

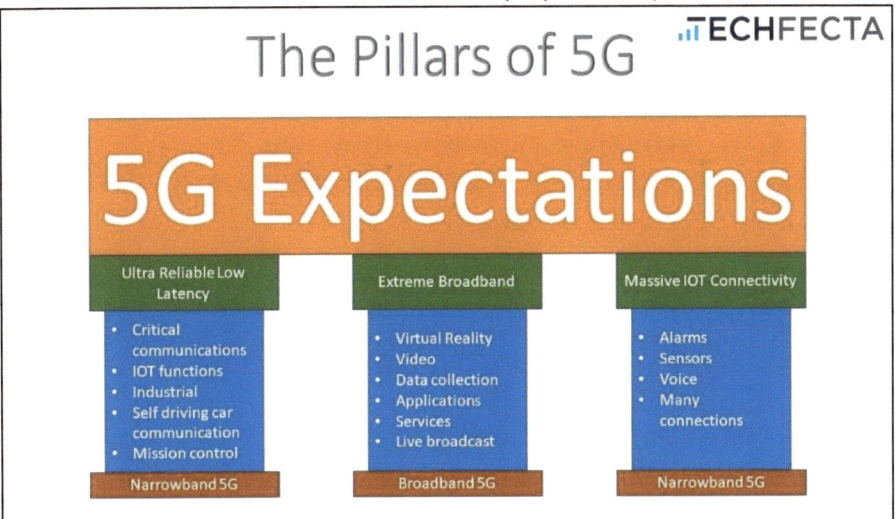

Figure 10 Pillars of 5G

Resources:
- http://kom.aau.dk/~nup/2016-06-27_Yilmaz-5G%20Ultra-reliable-Low-latency_final.pdf
- http://www.3gpp.org/news-events/3gpp-news/1774-5g_wiseharbour

Small Cell and CRAN Deployment Report

Summary:

Small cells are part of the ecosystem, and they are needed for offload and coverage fill. It's not going away. It's just not the mass deployment we had hoped for. However, as coverage and demand are needed, the small cell solution will be thought of in specific situations. It is a tool in the arsenal of weapons that the carriers use to deploy.

Outdoor small cell vendors:

- **Nokia** – OEM that makes indoor and outdoor small cells. They provide all kinds of small cells. Indoor and outdoor for all the carriers. They built the relationship off their macrocell business. It helps that they bought Alcatel-Lucent because ALU had a very strong small cell portfolio. Nokia had a great innovation when they created the mini macro. Some carriers use this in a way that most would not expect. They use it in rural and suburban areas. It is not big enough for larger coverage, but it's small enough to provide coverage in ears that don't need heavy loading. It is very limited in what it can do, like a larger small cell, but way short of what a macro site can do.
- **Ericsson** – OEM that makes small cells. They provide indoor and outdoor small cells. Ericsson reluctantly got into the small cell business. When small cells were introduced, they made the statement that they were going to concentrate on their core business of macro sites. They had to backpedal quickly because small cells were very popular.
- **SpiderCloud** – OEM for small cells and their target audience is the enterprise customers. They offer an indoor small cell that is approved by 2 major carriers in the USA.
- **Samsung** – OEM is another major supplier of macro and small cell. They have great products and supply all the major carrier's equipment. They have great products, and they are on the cutting edge of 5G. Traditionally they start strong, but they have trouble maintaining the cutting edge.
- **NEC** – OEM had small cells but didn't make a difference with most carriers in the USA.
- **Huawei** – OEM which is a powerhouse everywhere in the world but not in the USA where they are banned because of their close ties to the Chinese Government.

- **CommScope** – OEM has a great LET product. They acquired this product when they bought Airvana. However, I have not seen it deployed anywhere, yet.
- **ZTE** – OEM is another Chinese powerhouse, but I don't see their product anywhere in the USA.
- Airspan – OEM has built momentum by winning a major Sprint deployment with the Magic Box. This is a small cell breakthrough, it uses the UE relay as a backhaul and is a powerful
- **ip.access** – OEM
- **Ubiquity** – OEM

Resource

- https://www.thinksmallcell.com/Femtocell/femtocell-original-device-manufacturers.html

Increase Small Cell Value

Did you think that small cells should be more than radios? What if they could be the part of FOG computing? 5G asks that IOT has URLL, ultra-reliable low latency. How would that work? It's not that hard. We need to put an edge server in the small cell. I think this is being researched now, but the reality is that small cells are in a prime position to serve IOT markets.

Small cells are often looked at as 2 things. They can 1) fill the coverage gap or 2) offload the macro site. Now, small cells need to think beyond that, they need to offer more features, and they need to be more versatile.

If the small cell can serve as a FOG node, allowing the IOT devices to have extremely low latency by routing the traffic where it needs to go or responding immediately, then the small cell is more valuable in the system and satisfies a 5G requirement. This is where the small cell should be going. Small cells need to offer more than coverage. They need to be part of the actual 5G solution.

They also need to be creative. When I look at what Sprint and Airspan did with the "Magic Box" and using the LTE UE backhaul for the solution, that is amazing. They could make a cost-effective box that made the fronthaul accessible anywhere and used their LTE spectrum for backhaul, and it works.

Carriers were slower to deploy small cells for several reasons, but they all boil down to payback and value. They wanted a cost-effective solution for coverage, but after they got it, it wasn't enough, and the barriers to deploy were still high, fiber installation and monthly charges did not come down as they had hoped. Site acquisition and rent never got to where they wanted. The OEMs did their part, they provided a cheap unit that works like a cell site, but the other costs were prohibitive. That's one reason why some carriers would rather deploy CRAN because you have a sector that is control by a macro BBU or cloud controller in the size of a larger small cell. All the features for less money tied into a macro controller. It makes sense. While people see a small cell, it really isn't. CRAN could be a cloud or concentrated RAN. RAN is Radio Access Network. This is a standard solution that appears to be catching on with all carriers

Make the small cell more valuable. It should add more than coverage. Don't get me wrong, coverage still matters, but in today's world that is not enough! Especially when looking at the payback. If a small cell, especially an outdoor small cell where there are so many barriers, could add more value than coverage then it would be perceived as a high value-add product.

What small cells should have been multiple features. MIMO is a great start. Then a service for FOG computing features to take the loading from the cloud and core and put it at the very edge of the network. Then it should offer multiple wireless backhaul options, like carrier LTE, LTE-U, CBRS, and so on. If Sprint and Airspan could make the "Magic Box" work, then the OEMs should be able to make a unit that has more features.

Then, look at what the indoor small cells could offer if they were to take this to the next level. I am not saying they need to put a server with each small cell, that is not practical indoors, but they should have the option to tie to a local server for the edge computing so that it is more than a "hot spot" I mean you might as well deploy Wi-Fi if you just want hotspots, here we will have the LTE coverage throughout the building with edge computing to serve customers, devices, and IOT.

Let's not forget network slicing. Here is where the definition of MEC, Multi-Access Edge Computing. I know, you all thought it was Mobile Edge computing, but now, just to make things more confusing, they added Multi-Access Edge Computing so that you can get more network efficiency. It will allow each service on the network to work in its own realm to make each connection as efficient as possible. If you're in an office and an alarm goes off, you want to get that alarm right away, not after your coworker is finished watching 10 kitty videos on his new iPhone X, am I right?

My point here is that small cells need to keep moving forward and progressing. This is important to the carriers but also to the private LTE systems. At least if the FCC ever released any CBRS spectrum to us, ordinary folks. We want to build systems today. I mean at this point even T-Mobile should be a little frustrated because it should be delaying their densification plans.

Small cells could be a key component in IOT growth. The idea is that the small cell can use all these features to make IOT simple to roll out and connect to any device. It's about coverage and low latency. Here is an opportunity for the small cell to make a big difference in IOT deployment. Let's face it, LTE-M is a great solution for the IOT market. It is something that we could connect quickly and efficiently. Why not design the small cell to work efficiently with the IOT solution that will eventually merge into the 5G solution? Let's plan for this today!

Will Small Cells work with IOT and become the FOG edge?

As 5G takes off, there is more than just wireless over the air 5G NR (New Radio) format that is going to roll out. There is the entire network. One requirement will be URLL, Ultra Reliable Low Latency.

Could a small server be put in small cells to control IOT and act as a FOG server?

Could IOT feed small cell growth?

Make small cells part of the 5G solution.

It's all about the Value!

The value comes from the solution, the solution is determined by the application. What do you need? Where do you need it? How many people will use it at any given time? You've got to add value! The value helps to determine the payback. Then you have a system that makes sense, that hopefully, you won't have to replace in 2 years because it's insufficient. Also, a system which you are paying an arm and a leg for because the loading is so small that the system won't pay for itself.

Value is in the design, the use of the system. If you have the right system for the right application, then the payback makes sense. Let's plan accordingly. There are always surprises and limitations that could hamper the desired outcome, but we do the best we can with the knowledge we have. It's all we can ask for.

I hope this helps you bridge the gap between the spend versus payback!

This is great, but we still need fiber. I think if the small cell can connect to a macro site, then the fiber situation is resolved. The backhaul matters. An upcoming post will be about the backhaul, mid-haul, and fronthaul. I am a huge fan of a hybrid backhaul system which includes wireless and fiber because that is what will make the deployment of the 5G solution most efficient. New 5G spectrum will allow the fixed wireless solutions to be a game changer in this solution. More of that to come.

Resources:

- https://wade4wireless.com/2017/05/08/sprint-and-ue-backhaul-equals-a-magic-box/

- https://wade4wireless.com/2015/10/19/what-is-lte-ue-backhaul/
- https://wade4wireless.com/2018/02/26/how-do-we-get-more-backhaul-bandwidth/
- https://wade4wireless.com/2017/03/06/mounting-small-cells-in-a-city/
- https://wade4wireless.com/2016/10/11/deploying-5g-small-cells/
- https://wade4wireless.com/2016/01/04/small-cell-installation-checklist/
- https://wade4wireless.com/2015/07/27/small-cell-fronthaul-and-odas/
- https://wade4wireless.com/2015/11/12/wireless-deployment-handbook-for-lte-small-cells-and-das/
- https://www.sdxcentral.com/mec/definitions/what-multi-access-edge-computing-mec/
- https://www.comsoc.org/netmag/cfp/5g-ultra-reliable-low-latency-communications

Small Cell Installation Checklist

For all of you that want to do small cell installations for the carriers, this may help you along. Let's start with a high-level checklist to verify the steps before doing any installations.

Quick, high-level checklist:

- Get certified by the OEM
- Know the local permitting and ordinances
- Landlord issues
- Site Survey
- Outdoor is different from indoor
- Where is power coming from?
- Traffic control
- Grounding
- Mounting
- Testing and commissioning
- Optimization

Installation

Here is where the hardware gets installed. Let's look at what needs to be done.

The indoor small cell is usually 2 watts or less and could be 12 inches by 6 inches by 4 inches or less and generally will weigh 10 lbs. or less. They could have a very small and light antenna on them for either Omni directional coverage or directional coverage. Indoor small cells could remind you of the Wi-Fi access point, unit but bigger.

If you're installing an outdoor unit, then it may be bigger, and you must weatherproof the connections.

If it is a CRAN system, then it could be DAS, or you may have a DAS system that may have receive antennas or a front end with a BBU.

Remember that we are doing a quality installation, I have seen some horrible Wi-Fi installations. This is a quality installation that will work great and look very nice or be hidden from the customer, get it?

Small Cell and CRAN Deployment Report

Let's get ready to install.

By the way, Carriers need to Free the Small Cells!

Question Checklist:
- What are you installing?
    - Small Cell
    - CRAN
    - DAS unit
    - Backhaul
    - Fronthaul
    - Cable runs for fiber, copper, or RF
- Do you have the NTP, Notice to Proceed, from your customer?
- Do you have permission from the landlord or tenant? If your customer has it, then maybe you should have a copy.
- Is the landlord/tenant expecting you?
- Were you trained and certified by the OEM?
- Do you know how to ground the unit properly?
- Do you know how to clean and terminate fiber?
- Do you know how to crimp and terminate CAT5 and CAT6?
- Do you know how to mount the unit properly and permanently?
- Do you know the antenna alignment? Do you know how to connect the antenna and weatherproof it if needed?
- Did you document everything properly?
- Do you know how to commission and test the unit?
- Do you know the closeout package requirements of your customer, so you do not need to return to the site?
- Installation Preparation
- Make sure you have all your ducks lined up prior to the installation. I am talking about all the things that you may need on site.
- Did you make an appointment with the landlord or tenant or manager?
- Did you confirm?
- Do you have a name and number of the onsite contact?
- Do you have the equipment you are installing or are expecting prior to the installation? Can you track it to see if it was shipped?
- Did you record the serial number of the unit you are installing and document it?

- Do you have any certification that is required to show anyone on site in case you are asked?
- Did you or someone test the unit you are installing? Are you sure it will work?
- Do you have all the stuff you need like cables, caulk, connectors, hardware, and anything else needed?
- Did you review the site survey? Did it prepare you for the installation?
- If it's outside, do you need traffic control?

Will you be able to install at that particular time of day? A pole installation may require you to mount at night. While inside a building they may want you there at lunchtime when no one is in an office. Know your schedule ahead of time if possible.

Do you need to have a lease in place before mounting, is it in place?

Site Acquisition Checklists

I believe that many of you working out there are not sure what all site acquisition teams do. So, I thought by reaching out to some of them they could provide a checklist. Let me tell you that it is a tough job because there is so much out of your control. For instance, the lease and loading are up to the tower owner as to how to proceed. They aren't always going to move quickly to accommodate you unless you are a larger customer, or they are very slow. The lease involves legal people so that can slow everything down because there are so many "what-ifs" for everyone to hypothesize about. Then there is the permitting and zoning where you are at the mercy of the township or community which may only have a meeting once a month, and you may need to show up only to have your paperwork tables because Bob's hogs were loose in the neighbor's yard and they had to resolve his issue first. Don't know what I mean; then you are so lucky you didn't have to go to that meeting twice. Don't get me wrong, many townships and cities have a good system where you apply, submit drawings and then you wait for the approvals or changes. By the way, most townships know very little about what should be on a tower, they only know that they don't like them unless you have a public safety conscious council. They favor towers and wireless.

So where was I, oh yeah, you need to find the site, get the lease, and get all the approvals to get on the tower. Don't forget you may need to improve the structural quality of the tower, so that may take time and I know it takes money

Small Cell and CRAN Deployment Report

Let's get ready to install.

By the way, Carriers need to Free the Small Cells!

Question Checklist:

- What are you installing?
    - Small Cell
    - CRAN
    - DAS unit
    - Backhaul
    - Fronthaul
    - Cable runs for fiber, copper, or RF
- Do you have the NTP, Notice to Proceed, from your customer?
- Do you have permission from the landlord or tenant? If your customer has it, then maybe you should have a copy.
- Is the landlord/tenant expecting you?
- Were you trained and certified by the OEM?
- Do you know how to ground the unit properly?
- Do you know how to clean and terminate fiber?
- Do you know how to crimp and terminate CAT5 and CAT6?
- Do you know how to mount the unit properly and permanently?
- Do you know the antenna alignment? Do you know how to connect the antenna and weatherproof it if needed?
- Did you document everything properly?
- Do you know how to commission and test the unit?
- Do you know the closeout package requirements of your customer, so you do not need to return to the site?
- Installation Preparation
- Make sure you have all your ducks lined up prior to the installation. I am talking about all the things that you may need on site.
- Did you make an appointment with the landlord or tenant or manager?
- Did you confirm?
- Do you have a name and number of the onsite contact?
- Do you have the equipment you are installing or are expecting prior to the installation? Can you track it to see if it was shipped?
- Did you record the serial number of the unit you are installing and document it?

- Do you have any certification that is required to show anyone on site in case you are asked?
- Did you or someone test the unit you are installing? Are you sure it will work?
- Do you have all the stuff you need like cables, caulk, connectors, hardware, and anything else needed?
- Did you review the site survey? Did it prepare you for the installation?
- If it's outside, do you need traffic control?

Will you be able to install at that particular time of day? A pole installation may require you to mount at night. While inside a building they may want you there at lunchtime when no one is in an office. Know your schedule ahead of time if possible.

Do you need to have a lease in place before mounting, is it in place?

Site Acquisition Checklists

I believe that many of you working out there are not sure what all site acquisition teams do. So, I thought by reaching out to some of them they could provide a checklist. Let me tell you that it is a tough job because there is so much out of your control. For instance, the lease and loading are up to the tower owner as to how to proceed. They aren't always going to move quickly to accommodate you unless you are a larger customer, or they are very slow. The lease involves legal people so that can slow everything down because there are so many "what-ifs" for everyone to hypothesize about. Then there is the permitting and zoning where you are at the mercy of the township or community which may only have a meeting once a month, and you may need to show up only to have your paperwork tables because Bob's hogs were loose in the neighbor's yard and they had to resolve his issue first. Don't know what I mean; then you are so lucky you didn't have to go to that meeting twice. Don't get me wrong, many townships and cities have a good system where you apply, submit drawings and then you wait for the approvals or changes. By the way, most townships know very little about what should be on a tower, they only know that they don't like them unless you have a public safety conscious council. They favor towers and wireless.

So where was I, oh yeah, you need to find the site, get the lease, and get all the approvals to get on the tower. Don't forget you may need to improve the structural quality of the tower, so that may take time and I know it takes money for the engineering design, drawings, and work to be done. Then hopefully you can hang your equipment.

It's not like this for every build, just most of them. IT used to be fun to deploy, most towers could handle the load, now they are mostly loaded, and all the townships want to have a say in what you do.

Small cells should be easier, but they won't be. You will be doing more site acquisition at more sites. Hopefully, most of the poles you go on will be fine. If you choose to do rooftops, then you may still need permitting and zoning along with the lease. These need to have lease terms agreed to and signed. Even if you strand mount you need to pay someone to be there.

So, I thought this list would be something that could help you out. Here is the high-level checklist for site Acquisition. I have a detailed list, but it is long and full of detail. Let's start with the high level so you can digest it.

- Site Acquisition Firm:
- Site Acquisition Specialist:
- RF Engineer:
- Site ID:
- Candidate Name:
- Landlord:
- Candidate Address: (Nine-digit zip code required)
- Jurisdiction / County:
- Latitude: minute second' degree."
- Longitude: minute second' degree."
- Type of Structure:
- Existing Height:
- Height Available:
- Number of Carriers on the Structure:
- Will Structure need to be modified to hold equipment?
- Equipment Space Available:
- Zoning Classification / Process:
- Building Permit Process:
- Directions to Site:
- Property Owner & Parcel # required:
- Power info

Small Cell and CRAN Deployment Report

- Telco info
- No, this is not all of it. What about pictures? It is a good idea to gather as much information as you can. Is this a tremendous help or what?

Use Small Cells to Build a Private LTE Network

Build your own Private LTE Network

The great thing about 5G is that we will soon see private LTE networks. How is this possible? Because we finally have spectrum open to businesses everywhere. We already have license-free spectrum in 2.4GHz and 5.8GHz that we currently use for Wi-Fi. The FCC is going to allow users to use LTE in that spectrum. The issue is sharing the spectrum. It's not efficient in public places. It makes a big difference in your home. Suddenly when the devices have LTE, you can install a private LTE license free network in your home. That is cool.

Now, imagine that you can build a licensed LTE network for your small business or to serve your IOT needs to for very high security on your device? You could have your device registered with a SAS company to coordinate your CBRS spectrum so that it's your spectrum in that building, office park, or wherever. You could even use it on your smartphones when they add that spectrum to their RF boards. This is expected to happen in 2018.

Why Private LTE?

If you want a fast and reliable network, then a private LTE network is the way to go. You could improve security by going to a licensed carrier. That is why the CBRS spectrum will be a good fit. You can easily grab a lightly licensed spectrum for your personal use, and it will make your network very secure. If you need the speed and reliability that Wi-Fi may not provide, then this is a good alternative.

Why would I want a private LTE system?

The big thing now is the industrial IOT functions. This is where you may have a manufacturing plant or a warehouse where the latency and reliability are critical. This is an IOT function where IOT would make a difference, and it could change the way your machines communicate. It would be dedicated to your specific purpose.

I would like to say that your devices would have it but that is about a year away. The latest iPhone did not have this spectrum in it nor did it have the 600MHz spectrum that T-Mobile is building out now. iPhone appears behind the latest technology spectrum.

However, someday all devices will have the CBRS spectrum in them, and your device can hand off to your secure internal network and then back to the carrier's network, in theory anyway. The idea is that we can do so much more with our devices that could be dedicated to our specific business. That is the dream that we can run our specific applications that matter to our business. Let's say on our tablets. If you want a model, look at any scanner system for inventory, they use this now to scan everything. Imagine when we can put applications that require more bandwidth on smaller devices and take them anywhere. Our smartphone is like that now but on the carrier's network and Wi-Fi.

Wi-Fi serves a great purpose, but security has been the issue. If we can dedicate a specific channel in the CBRS to a specific function on your smartphone, it will be very secure. It all depends on what your priority will be to achieve your networking goal. Is it easy access, security, functionality, or all 3?

What is the CBRS?

The citizens broadband Radio spectrum. This is currently specific to the United States, but it is going to open everywhere if it's successful. US CBRS is the 3.5GHz band, which runs from 3550 to 3700 MHz band. CBRS stands for Citizens Broadband Radio Service (in remembrance of the CB, Citizens Band). It is a licensed spectrum. There is Military radar, and Earth stations that use this spectrum that is grandfathered in and have priority access. That will not change. There will be a Spectrum Allocation System (SAS). Currently in the US only, but Europe is looking to follow suit with Licensed Shared Access, (LSA).

The spectrum is 3.65GHz to 3.7GHz which was used for WiMAX. Now the FCC is opening that spectrum and an addition 150MHz spectrum for 3 types of users.

What the SAS will manage:
- incumbent access including the federal government and satellite providers.
- priority access licenses (PAL) which are 7 10MHz licenses to be awarded to the highest bidders. PALs will be protected from the GAA users. PAL will include commercial users like carriers, rural operators, are a 3-year license with only 1 renewal term allowed now and will be in the 3500 to 3650 portion of the spectrum. One licensee can hold only 4 PAL licenses.
- general access user, (GAA) which is "Licensed by rule" which requires the rules to be followed. This will be dedicated in the 3650 to 3750 MHz portion of the band.
- A PAL may gain additional GAA spectrum.

- Companies that currently have this spectrum licenses will be able to keep their licenses. This was used for WiMAX in the past. Now it will be LTE focused.
- Licensing will be done by the Spectrum Allocation System, (SAS), which is a group that can charge for these services, currently being led by Google and Federated Wireless.
- Hardware vendors include SpiderCloud, Ruckus, Nokia, Ericsson, Samsung, ip.access, and Acceleron.

The 3 tiers are:

- **Incumbent access** – this is for users that are already using this spectrum for the military, ground stations, government, and so on. They will be protected.
- **Priority Access Licenses (PAL)** – this is for anyone who is willing to pay a premium to own 10 MHz of spectrum. The current model is for 3 years with one renewal, that could change.
- **General Authorized Access (GAA)** – this is for anyone who wants to use it if they have an authorized device that will connect to the SAS, Spectrum Allocation Service, and accept the assigned frequencies. You must complete a questionnaire and pay a small subscription fee, but it's going to very reasonable.

The key here is that the SAS will coordinate all users, protect the PAL users and the incumbents.

No PAL users will be on the lower spectrum, and no GAA will be in the upper spectrum.

PAL users will most likely be the carriers or anyone who is willing to pay for dedicated spectrum for data applications, like broadband, IOT, VOIP, or anything that could be a wide area densification project.

GAA users could be for anyone in a building doing any type of LTE network. This could be private and secure coverage in a building or IOT applications or manufacturing applications that require very low latency. A private LTE network on a lightly licensed network.

A PAL can grab GAA spectrum but a GAA can't grab PAL spectrum.

While the GAA users should not interfere with each other, it could happen, not much you can do.

All users need to comply with the FCC rules.

CBRS and the Shift in Spectrum Ownership

I have been speaking about how the spectrum of 5G will shift into the hands of the small business once again. Well, now there are more people on board with this theory. It seems that CBRS is making it all happen, and IWCE had CBRS as one of its focal points. (Even though I could not make it this year, they talked about it!)

Quick update, the US CBRS is the 3.5GHz band, which runs from 3550 to 3700 MHz band. CBRS stands for Citizens Broadband Radio Service (in remembrance of the CB, Citizens Band). It is a licensed spectrum, but it is split up into 2 areas. There is Military radar, and Earth stations that use this spectrum that is grandfathered in and have priority access. That will not change. There will be Authorized Shared Access, (ASA). Currently in the US only, but Europe is looking to follow suit with Licensed Shared Access, (LSA).

ASA includes:

- Incumbent access including the federal government and satellite providers.
- Priority access licenses (PAL) which are 7 10MHz licenses to be awarded to the highest bidders. PALs will be protected from the GAA users. PAL will include commercial users like carriers, rural operators, are a 3-year license with only 1 renewal term allowed now and will be in the 3,500 to 3,650 MHz portion of the spectrum. One licensee can hold only 4 PAL licenses.
- General access user, (GAA) which is "Licensed by rule" which requires the rules to be followed. This will be dedicated in the 3650 to 3750 MHz portion of the band.
- A PAL may gain additional GAA spectrum.
- Companies that currently have this spectrum licenses will be able to keep their licenses; this was used for WiMAX in the past, now it will be LTE focused.

Small Cell and CRAN Deployment Report

- Licensing will be done by the Spectrum Allocation System, (SAS), which is a group that can charge for these services, currently being led by Google and Federated Wireless.
- Hardware vendors include SpiderCloud, Ruckus, Nokia, Ericsson, Samsung, ip.access, and Acceleron.

I have been continuously explaining how CBRS will become a major player if the vendors pick up on it. Well, Google seems very interested. My friend Tom Ulrich put together the following report that covers how Google is excited to work with this spectrum and how it is the new beachfront property. How it will open new doors for all of us to deploy over the next 5 years or so.

IWCE had entire workshops on it where all the big OEMs were there to promote the spectrum. They see great opportunity for growth here. One such workshop was "Building an Ecosystem for the CBRS Band" that had all the big players there. Nokia, Ericsson, Ruckus, Google, WISPA, Airspan, Federated Wireless, Comsearch, Telrad, and Cambium Networks were all presenting something about what they could do to contribute. They all see great potential in this. If you are a system integrator or do network implementation, then hopefully you see the potential as well.

I was reading a report by ABI Research that mentions several articles pointing to Verizon Wireless plans to use CBRS to replace middle price DAS systems, the articles in RCR and Fierce Wireless using CBRS as the neutral host solution. Then it shows how Nokia added the CBRS to its Airscale product and the Ruckus OpenG product to follow suit. Not to be outdone but Acceleron also has a CBRS product. Just to be fair, Spidercloud was one of the first to have a CBRS product. It spears that Spidercloud is already reaching out to the DAS vendors and Verizon to bridge the gap for smaller DAS systems. We shall see more of SAS, (Small Cell Antenna Systems) popping up to replace the smaller DAS systems.

Could CBRS solve the DAS middleware problem? Could CBRS products fill the void where no one wants to invest in those 100,000 to 500,000 square feet venues where it is too small for a carrier but too large for a small cell? Is this the savior we are looking for? I hope so! A clean way to hand off and a lightly licensed spectrum where we would not all be trampling on each other in the Wi-Fi space. I see a solution that could solve so many issues, financial and technical.

While this will mostly be an indoor solution, something where we could replace some DAS system with a common platform licensed spectrum that all the carriers and non-carriers could share to reach the dense population, it will be used for enterprise and outdoor coverage as a critical part of the 5G network slice. I am looking forward to seeing what small businesses can do with this spectrum to serve the people.

If you want a quick overview, here are 2 links that can help:

- https://www.fcc.gov/rulemaking/12-354
- https://www.thinksmallcell.com/LTE/what-is-cbrs-shared-spectrum-for-in-building-small-cell-wireless.html

I focus mostly on the enterprise play here, but the reality is that we can use this spectrum for more than just indoor solutions. I see the spectrum to be used for new solutions like backhaul in the tough area or as a fixed wireless solution placed where we need limited spectrum over short distances. I also see the carriers using this as a common small cell solution that can handoff from the licensed LTE spectrum we see today to be used to fill small holes without the very expensive LTE spectrum that they FCC auctioned off for a very high price. I see cost-effective small cells in public area where the more expensive solutions from, the bigger OEMs are not practical. Price matters, but the high cost of backhaul is one of the limitations that hold back deployment, along with permitting costs. All of this is restricting small cell deployment today causing the FCC to push legislation to streamline coverage. Everyone wants great coverage and high bandwidth, no one wants to see an ugly tower in his or her backyard.

I see CBRS filling the public venues with an alternative to smaller DAS systems by dropping in a CBRS small cell with multiple bands to provide a lightly licensed signal where the carriers would roam onto this device. Clean signal without the threat of another access point going up on the same band(s).

CBRS will allow small business and Enterprise to have their lightly licensed spectrum, something that the FCC has kept from small business for quite some time. I get it, they make billions on the auctions, but it has not helped small business broadband. They feel the ISM band was enough for them to build on. I feel differently. Now I see opportunity in CBRS, centimeter wave and millimeter wave spectrums. Let's deploy and bring broadband and narrowband to the masses! Broadband for internet access and narrowband for IOT access. It's exciting to see the industry have more opportunities again!

Small Cell and CRAN Deployment Report

Tom's report put together some notes from IWCE. Here is Tom's report from IWCE, *Is CBRS ending "Beachfront Spectrum"?*

I had the pleasure to attend IWCE this week and was blown away by Dr. Preston Marshall's {Alphabet/ Google} presentation on CBRS.

If I had to describe his presentation in 3 words, it would be:

- *Ecosystem*
- *Incentive*
- *Innovation*

Is the Ecosystem of Spectrum Landscape changing? Will CBRS end the need for Beachfront Spectrum?

It is first important to look closely at the "Current reality" of Wireless Spectrum: How does the current Spectrum landscape preclude innovation?

Licensed Spectrum is Expensive – Past Auctions cost the WSP's Billions to own the right to this FCC Licensed Spectrum and only come available once every ~3-5 years. They also really limit the number of participant and winners. Do you have a Billion dollars to purchase spectrum for your "Garage idea of the Century?

Newly Licensed Spectrum rollouts are meticulously planned and take forever to plan/ execute. They often force the WSP's to commit to the next technology type before knowing how successfully adopted it will be. Even after the Spectrum purchase, look at how many Billions of Dollars were committed to development/ deployment in WiMAX for Intel, Google, & Sprint before changing to an LTE-based solution. Look at how difficult turning off old technology types {Analog, iDEN, GSM, UMTS, & CDMA} have become.

Spectrum is Slow to deploy and can take 6-8+ years to clear spectrum, raise funding, and establish a product rollout. Look at failed Spectrum rollouts like LightSquared, Next-Wave, etc. Some companies like Dish have even had the forward thinking through of saving spectrum, waiting for the next technology shift, or WSP Spectrum shortage to capitalize on their dormant Wireless portfolio.

In today's unlicensed wireless ecosystem, it encourages OEM's to make cheap, lousy radios that do not perform very well with interference present. 802.11

Wireless AP's are often a cheap commodity that needs to be upgraded or replaced every 3-5 years. How much innovation can we drive with a $40 access point? This often drives the race to the bottom on who can create the least cost AP.

How does CBRS set-up to change the Spectrum Landscape and Drive Innovation?

Dr. Marshall stated, "CBRS will make spectrum buying an economic decision." It incentivizes stakeholders to maximize their ability to deflect interference and operate with radios that can perform in a noisier shared spectrum environment.

Dr. Marshall detailed his 4-step plan to roll out CBRS

1. *Regulatory – Helped get FCC approval, help develop standards within the Wireless Innovation Forum, and CBRS Alliance. Established FCC Part 96.*
2. *Coexist – Creating an ecosystem environment for Multiple Technologies and Stakeholders*
3. *Recruit – Recruit top talent and buy-in from Wireless industry*
4. *Prove – Further innovate standards, product, Solutions, and Applications.*

CBRS creates a Wireless Ecosystem that now will encourage innovation and allows for fast, less expensive rollouts. Why not put a solution in the Marketplace and let the market decide how well it is adopted before committing to extensive field trials and Millions of dollars?

Dr. Marshall detailed that this Spectrum Landscape is sustainable to support additional shared spectrum bands and may hold some of the keys on Business models, and landscape of 5G.

Special thanks to Tom Ulrich!

Now, in my opinion. We have seen the players be OEMs and carriers and other integrators in this space. Who has been conspicuously absent has been the cable companies. Here is space where they can shine, grow, and spread beyond Wi-Fi without building an ironclad agreement with one carrier. They have the money and the deployment process to make this a phenomenal area of growth. I would like to think that SpiderCloud would be calling the cable companies with proposals and business cases. Just my opinion. It is time for the cable companies to make it happen in wireless deployment.

Your Private LTE Network

You can use the licensed or unlicensed spectrum to build your own private LTE network. This will be part of the 5G ecosystem. It could be a separate network slice, from my perspective.

You will need:

- A mini-core to control your systems and to be the interface to the internet.
- Radios that are in the band and spread throughout your little area.
- What will they control? Devices, smartphones, laptops? What will they connect?
- User Equipment which is the end user's device. It could be a card to interface with your device if doing IOT. It could be your smartphone, which should have this spectrum in it starting in 2018. Maybe your laptop will be able to connect or have a USB interface that could connect.

Now you see that there is a way to get that private LTE network, but where do you get the parts for a CBRS network? Look at the list below:

- Ruckus makes radios and a small core.
- SpiderCloud also makes radios and a small core.
- Look to Federated Wireless to see that they already have run trials, http://www.federatedwireless.com/tag/3-5-ghz/
- Any major OEM has the equipment, like Nokia, Ericsson, or Samsung, but they generally have little interest in helping a smaller business with something like that. That's been my experience.
- As for end-user devices, I am still trying to figure that one out.

There you go, figure it out. It's so easy a wireless guy can do it. Maybe an IT guy can do it, but who knows.

Resources:

- https://wade4wireless.com/2017/07/25/art-king-teaches-cbrs/
- https://wade4wireless.com/2017/07/31/cbrs-deep-dive-with-steve-martin/
- https://wade4wireless.com/2017/04/17/cbrs-and-the-shift-in-

spectrum-ownership/
- https://wade4wireless.com/2016/04/12/cbrs-citizens-broadband-radio-service-update/
- https://blogs.cisco.com/sp/wait-for-it-wait-for-it-5g-its-here
- http://spidercloud.com/cbrs
- https://www.cbrsalliance.org/
- https://www.fcc.gov/rulemaking/12-354
- https://www.thinksmallcell.com/LTE/what-is-cbrs-shared-spectrum-for-in-building-small-cell-wireless.html
- https://www.atis.org/wsts/docs/2016/4-01_Google_Peroulas_Freq_Time_Phase_CBRS.pdf
- https://www.waterfordconsultants.com/reference-materials/CBRS_Spectrum.pdf
- https://www.qualcomm.com/invention/technologies/lte/private-lte-network
- http://www.privatemobilenetworks.com/solutions/4g-lte/
- https://www.networkworld.com/article/3179784/mobile-wireless/the-big-cbrs-promise-private-enterprise-lte-wireless-networks.html

Small Cell and CRAN Deployment Report

Small Cells in LAA, CBRS, LTE-U are 5G Building Blocks!

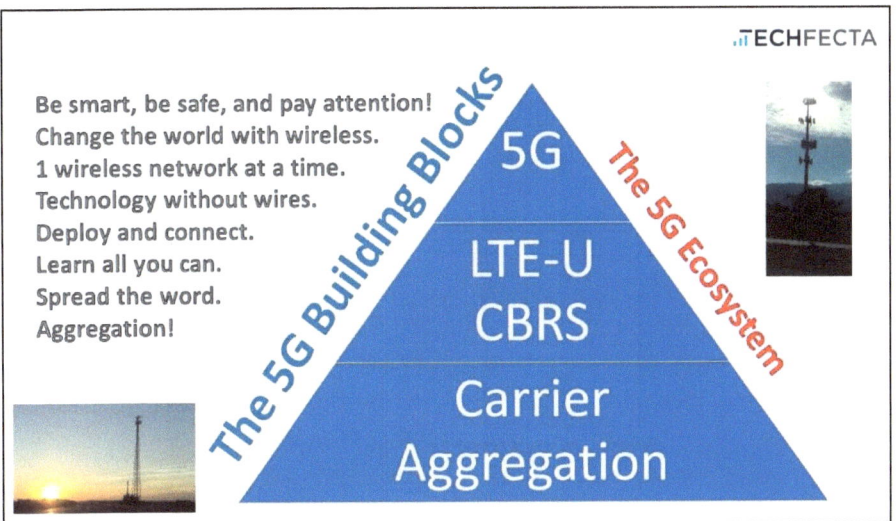

Figure 11 5G Building Block, CA, LTE-U, CBRS

Carrier Aggregation and Private LTE

Hey, guess what? Remember when I talked about LAA, LTE-U, and CBRS, well, we're almost there! That's right, all those add-ons that we talked about over a year ago may finally be released. The new CBRS spectrum in the US creates new opportunities. It will be a gamechanger once the FCC decides what will happen with the spectrum. I just want to deploy, something they are delaying even longer, over a year, come on, release something! Sorry, it's just that government holds back the economy by trying to improve it, this is a classic example. It happens in every administration.

Let's look at 2 things, 1) a way to supplement what the carriers have, and, 2) a private LTE network. While you think this is just another system, it's on the road to 5G. That's right, we need spectrum, and there is plenty of it. Where? I'll tell you if you have the nerve to read on.

LAA – Licensed Assisted Access

First, it's the ISM spectrum, you know it better as the Wi-Fi spectrum. It could be 900MHz, 2.4GHz, 5.8GHz or even 60GHz. All that spectrum that Wi-Fi could operate in, but they primarily use 2.4GHz and 5.8GHz. Now we can run LTE in that spectrum; it's called LTE-U and Qualcomm created MuLTEfire to allow LTE to run without a core. You heard me, an independent LTE-U hotspot that the carriers and your device will like a lot.

LTE-U is in the spectrum of Wi-Fi but is true LTE. It's in the unlicensed band. MuLTEfire is a stand-alone LTE unit. While that doesn't mean much now it will when all the devices have it, like your smartphone. Learn more here, https://www.multefire.org/about/ if you're interested in learning more. I see this as a great thing. RCR wrote about it last year, https://www.rcrwireless.com/20170530/network-infrastructure/what-is-multefire-tag6-tag99. It's really a great thing.

Why? I am glad you asked. You see the carriers can use LTE-U to gain bandwidth with their current LTE spectrum. They use something called LAA, which I wrote about in the past, but now it's right around the corner. What's that? You don't remember anything about LAA, let's have a flashback! LAA, which is soon to be used by the primary carriers to build up throughput, is Licensed Assist Access. That uses carrier aggregation to aggregate, (make all the RF carriers look like one big pipe), create separate streams that all combine in the end users' device to make the throughput go through the roof. We are talking about throughput closer to 1Gbps to the device. Hell, I would settle for 100Mbps to happen.

There is also LWA, Licensed Wi-Fi Aggregation, same as the above but it uses Wi-Fi to aggregate. Now, you're wondering why LWA isn't the preferred choice, right? I would say because it's so much work to get Wi-Fi to work with LTE. Seriously, those 2 formats are not playing well together. They have delay differences and to be honest the carriers are sick and tired of trying to make it work. I will give T-Mobile a lot of credit because they seemed to have a solid way to make it work. I am not sure why, but they don't seem to push it anymore, not sure why.

So, what is the better choice? LAA using the licensed LTE with the unlicensed LTE-U. LTE is a great format baby! I love it, in fact, 5G will love it too, it will be the foundation for 5G, even with the NR coming out. That's right! The carriers will be rolling out LAA very soon. They have faith in it. Carrier aggregation rocks! It will allow all of them to unify LTE is all bands, including CBRS.

CBRS – Citizen's Broadband Radio System

That's right, CBRS rocks! It will allow you, the end user, to build a lightly licensed spectrum, like 3.5GHz here in the USA, and build your own LTE network along with the LTE-U. CBRS is in the 3.5GHz spectrum, and it is lightly licensed thanks to an advances licensing system that uses ASA to assign spectrum. It works a lot like your DHCP server only for spectrum. How cool is that?

When I interviewed both Art King (of SpiderCloud) and Steve Martin (of Ruckus) they helped me understand how awesome the CBRS spectrum is going to be for the enterprise user as well as the wireless internet service provider, (WISP). It will open new doors and allow WISPs everywhere to align with the carriers if they want to aggregate the carrier.

But wait, there's more! The carriers will be using this spectrum as well. In fact, T-Mobile is a major reason the delay is happening. They want to buy more of the spectrum and hold onto it much longer. I'm okay with that, but why don't they ask the FCC to get off their ass and release something already? They don't seem to be in a hurry. I guess because they are rolling out 600MHz now. That is a major effort on their part. I appreciate it because it brings a lot of work to the deployment industry, and GOD knows we need it. You heard me! It's feast or famine in this industry. Now I am straying off topic, sorry.

CBRS is also LTE, and it's something that the carrier separately wants to roll out as part of their small cell model. It's perfect spectrum for small cells. Lower power, smaller coverage area, and new spectrum that can be rolled out anywhere. I love it and so will you when you get your own small cell in the CBRS spectrum.

Hey, guess who else will love it? Public safety, utilities, small business, enterprise, and anyone who wants to extend their coverage. I know, Wi-Fi is nice, but the LTE format will open new doors, and lightly licensed spectrum will help security. I can't wait until every smart device has this spectrum in it.

Carrier Aggregation made this possible!

I would say thank you to the OEMs for coming out with carrier aggregation. It is the sole feature that made this possible. It is a remarkable technology that the carriers are currently using only in the licensed spectrum. They are currently using this technology only on their own licensed channels. Some of them on multiple bands across spectrum or maybe in the same spectrum. Here's the deal, they can expand beyond their own spectrum, how cool is that. They may want players, so keep your ear open, and maybe they will want partners. It could happen.

Small Cell and CRAN Deployment Report

What about the devices?

Nokia has already released their CBRS small cells. Ruckus has a CBRS unit. SpiderCloud has their unit. There you go, real units in the real world just waiting for you to deploy! Deploy, deploy, deploy! Do it ASAP. We're still waiting on more UE devices to get the spectrum, but it's coming, have faith, my friends.

How can this help you?

OK, it's all about you! I can explain it to the industry you're in. Let's look below at the list to understand where you fit in.

- **Enterprise users** – you will be able to build something much more secure than Wi-Fi. A network in the CBRS will be lightly licensed and something that most users will not have access to. You could dedicate it to your network and have specific devices with that spectrum to allow low latency and dedicated spectrum for whatever you want to use it for. Security and dedicated spectrum would allow you to do secure functions without running cables everywhere.
- **ISPs** – If you're an internet service provider you have an option to supplement your income by striking a deal with any carrier to use LAA on LTE-U or CBRS to aggregate the carriers licensed LTE with your spectrum. If you have already built it, then they may see value in what you have. Why not?
- **Installers** – if you have the opportunity then you can install these devices indoors and out. On poles usually, they are low power and will reach the end user much like Wi-Fi on these will be small cells. The CBRS will be rolling out everywhere. While the carriers want a higher power unit the reality is this will be an extension of the network and should work well within the realm of 5G.
- **Backhaul** – Fiber providers, backhaul providers, and the router companies can make a play to gain market share in the smaller business groups.
- **Carriers** – the carriers will look at this as the icing on the cake. They can extend coverage by partnering with smaller users for a flat fee or by reciprocating services. The chances are good that they will want to build their own smaller networks where they can, but that whole idea here is that they don't want to increase CapEx. That's why the partnership to extend coverage looks natural and attractive. FYI – this is cheaper than putting in licensed radio heads.

Small Cell and CRAN Deployment Report

- **Building and venue owners** – if you looked at DAS as the saving grace, look at this as an easier way to do upgrades. This is an easy way to expand coverage by running CAT 5 or CAT 6 to the new small cell areas. By putting in LTE-U or CBRS, you can quickly and easily extend coverage. It's a natural alternative to putting in more licensed radio heads.
- **OEMs and distributors** – you can move product. Nokia, SpiderCloud, and Ruckus have already jumped on the bandwagon with products for LTE-U and CBRS that are carrier grade. They all are targeting the user mentioned in this report. Do they know something you don't? Not if you read the blogs at www.Wade4Wireless.com!
- **Public Safety** – here is a market that could really benefit from this. I know they are all waiting for the AT&T deployment, sorry, I meant to say FirstNet roll out. The thing is, they will have a choice between AT&T, sorry, I mean FirstNet and Verizon's system. It all depends where they are at because the coverage varies that much and while the budget matters, they really need to have coverage first. If the coverage is there, then they can look at cost. They may rely on their own networks as they do today. If that is the case, they may want to try to deploy their own broadband network and put it where they need it.
- **Utilities** – I see the utility market needing this for their smart meters and remote connections. They can use something like this to reach all the area that the carriers can't or at least as an alternative to paying carriers for their IOT systems. Then they may be able to expand the network to serve more than the meters by allowing end users to put their IOT devices on these networks. It is a viable alternative.
- **Cable companies** – I see the cable companies rolling out LTE-U and CBRS because it's so cost-effective and it fits into their current model. Granted, they love Wi-Fi, but they will see great value in rolling out LTE-U and CBRS if they can partner with the carriers. They could use it as a bargaining chip when they want to become an MVNO, like Comcast, with the large carriers. What leverage they will have when they can use their network to reach millions of their customers with an LTE network of their own. Awesome!
- **Autonomous vehicles** – I don't know who will use this, but it is a viable way to communicate with vehicles in the urban areas. It would have to be a dense network, but it could supplement the carrier's network.

- **Google** – yes, Google has a lot to gain with their Loon network and to replace the FTTH network which we all thought would be our savior from the choices we have now. One thing that Google learned is that it costs a lot of money to deploy fiber. It's a lot more than what you see in the data center. If they use this technology like they say they will, then we will have our new ISP competitor that could make a difference overnight, well, in a few years maybe.
- **The workforce** – I see the workforce benefiting from this because we are going to have to engineer and deploy the systems for all the potential service providers. Let's build the best networks we can. Let's build the ecosystem beyond the carriers to the smaller business owners. We can make a difference by helping cities become smart, utilities being connected, public safety has a broader reach and more partners, and the carriers are relying on individuals beyond their strong list of partners. We can go beyond the "norm" and into an open distribution system. We can do it, so let's make the difference today!

Summary:

Do you believe that this could be a game changer? I do! I think that this is going to be the thing that pushed the outer users to 5G and includes them as part of the ecosystem. It's more than being a customer. It's becoming a player in your niche. Finally, in wireless, we can all be players in the infrastructure network. We have become an active user and builder of the wireless networks beyond Wi-Fi. It's exhilarating to me, and I look forward to having all of you as partners in this venture!

Resources:

- https://wade4wireless.com/2017/07/25/art-king-teaches-cbrs/
- https://wade4wireless.com/2017/07/31/cbrs-deep-dive-with-steve-martin/
- https://wade4wireless.com/2015/09/08/lwa-laa-lte-u-and-wi-fi/
- https://wade4wireless.com/2017/12/04/what-is-carrier-aggregation/
- https://wade4wireless.com/2016/02/01/unlicensed-lte-multefire-overview/
- https://wade4wireless.com/2015/08/10/wi-fi-lte-lte-u-and-aggregation/
- https://wade4wireless.com/2015/06/29/small-cell-deployment-prep-notes/

Small Cell and CRAN Deployment Report

- https://www.thinksmallcell.com/Technology/what-is-laa-and-how-does-it-affect-small-cells.html
- https://wade4wireless.com/2017/09/26/building-your-own-private-lte-network/
- https://networks.nokia.com/products/td-lte/CBRS
- https://en.wikipedia.org/wiki/Multefire
- http://www.federatedwireless.com/tag/cbrs/

Indoor Coverage Matters!

I know what you're thinking, indoor connectivity, why does that matter for the city? I mean really Wade, who cares? Indoor connectivity will matter just as much as outdoor connectivity. After all, the smartphone should not stop working when you go into a building. If your phone stops working, how does that make you feel? It sucks, right! What about public safety, they can't have their radios die the minute they go into a building, that could mean death, literally, for them or whoever needs help. Indoor connectivity should be thought of as crucial for any city, much less a smart city.

Do you stop using your device when you go inside any building? Seriously? Of course not, you don't expect it too.

Indoor coverage included the entire building. It's sad when you see a disaster happen and people don't have coverage in the obvious places in a disaster, like the stairwells or the closets or basements. Why is that? Because normally people don't go there, and the building owner didn't want to pay for something that no one would normally use. That is why there are regulations, in some cities, to make sure that at least the emergency bands used by fire and police are working in those areas. The local fire departments and radio shops can put this in and test it. However, most business people don't enforce this because they know that businesses and building owners do not want to pay for it. We all think it's OK until a fire or a terrorist attack happens, and the people inside can't communicate because they are in dead zones. In an emergency, a dead zone could mean that the people could die because they could not reach help. That often gets overlooked just to save a few dollars in many cities.

So, when planning a smart city, the regulations matter, the rules to define whether a company needs to have the best coverage, wired and wireless, in their building really matters in the grand scheme. Why not think it through and look at what has happened in the past. Take the necessary measures to ensure that buildings are being built to the proper code for structure and safety.

How will we cover inside?

A lot like we do now, but we also want wireless coverage. Let's look at all the indoor connectivity options.

Small Cell and CRAN Deployment Report

Wi-Fi

This is the obvious thing that we all expect to see everywhere. I don't think I need to cover this issue because almost every public area has Wi-Fi and most offices have it as well. We expect to have Wi-Fi everywhere. It's an expectation in every office we work at.

In shared areas, we expect Wi-Fi coverage. The thing is, who will pay for it? If you go to an airport, it's nice when the airport provides it. When we must pay a Wi-Fi service provider, we aren't as happy. It's a hassle when all you want to do is download your boarding pass, and you need to shut off Wi-Fi to go back on LTE just to get through security or on the plane. However, that's what I do.

LTE-U

This is going to be on the same spectrum as Wi-Fi and the carriers are excited because the handoff from licensed LTE spectrum to unlicensed LTE is almost seamless if it's set up properly. This will be a game changer for all the carriers to share loading with devices in the ISM license-free band. WOW! A way for your smartphone to hand off its data and VoLTE, (Voice over LTE) calls to a spectrum that should not cost you any data on your plan.

If this can be put on every device, I would see it really is a game changer for the carriers to hand off to almost any vendors units. With the coming of age of the cloud and mobile edge computing, MEC, we will see things improve greatly.

DAS systems

DAS systems are still booming in high density and high traffic areas. They are being upgraded. While small cells are making a dent, they are being used together to provide better coverage for less cost.

You see, the original DAS systems could include a Macro site to feed it for the system to reach all the areas of the building, stadium, or whatever you're trying to cover. Now they can feed it with small cells. Now they can transport the signals digitally, meaning that instead of coax cable they can run fiber and use power from a local connection point or even run things through router and power the radio head with PoE, (Power Over Ethernet) which is really a great way to deploy.

DAS, (Distributed Antenna System), is a great way to get the signal out to the people, but it's a financial commitment that small and some mid-size business don't want to pay for. The carriers no longer see a payback on these systems, and they are looking for a less expensive way to get the signal out to the people.

Hey, I love DAS systems. They are crucial for the wireless infrastructure to cover venues. But the carriers are looking for more cost-effective ways to get the signal out. Now that we entered the age of a seamless digital network using LTE for wireless we can distribute the signal using fiber and CAT5 cable instead of the coaxial cable and splitters and analog amplifiers that we relied on in the past. DAS systems have evolved and will continue to evolve to meet the needs of today's market.

Public safety bands
This is critical, but it's an unknown. I don't talk about public safety bands much because it's going to go through changes. These departments still need to have the urgent PTT, (Push to Talk), Voice access because it's reliable and immediate. We don't want to wait for the emergency responders to be able to communicate in an emergency. Currently, they still rely on DAS and outdoor coverage.

They still need data to work on their laptops and gather information. FirstNet is taking care of this with their recent partnership with AT&T to provide dedicated coverage for first responders. I am waiting to see how this plays out. I am glad that first responders will have a priority channel, but most of them already have smartphones. They don't all have devices paid for by their jobs, many use their personal devices and coverage plans because not all governments have the budget to supply everyone with what they need.

For the emergency responders, there are rules in many areas that require buildings to put in DAS systems or radios so that emergency responders have coverage as I said before. I have no idea who has this requirement and who doesn't. It really seems to vary, even within cities.

While the public safety aspect will weigh heavily in the smart city planning, it should be thought of as part of the wireless and wireline deployment. Please don't make it an afterthought. It will take regulations to ensure that all indoor coverage is thought out and planned properly.

CBRS
There is an entire CBRS section above, but here are some more notes. The US CBRS is the 3.5GHz band, which runs from 3550 to 3700 MHz band. CBRS stands

for Citizens Broadband Radio Service (I remember the CB, Citizens Band, here in the US). It is a lightly licensed spectrum, but it is split up into 2 areas. There is Military radar, and Earth stations that use this spectrum that is grandfathered in and have priority access. That will not change. There will be Authorized Shared Access, (ASA). Currently in the US only, but Europe is looking to follow suit with Licensed Shared Access, (LSA).

This will mostly be an indoor solution, something where we could replace some DAS system with a common platform licensed spectrum that all the carriers and non-carriers could share to reach the dense population, it will be used for enterprise and outdoor coverage as a critical part of the 5G network slice. I am looking forward to seeing what small businesses can do with this spectrum to serve the people.

Factories have connected systems that may not be so reliant on Wi-Fi. Now they can dedicate a specific carrier to that function inside the factory so that no one may share it. Keep that spectrum dedicated to the machines and very low latency so that no one else can use it, jump on it, or break it. That is a game changer for indoor wireless!

I often overlook the use of indoor wireless for factories and distribution warehouses. This is a great use case for indoor wireless and one that needs low latency as well as dedicated spectrum. You want to keep it secure and dedicated for one purpose. Here is a perfect spectrum that they can add to any machine in their system. It helps to cut down on issues due to latency. Distribution will need to provide accurate order filling, and factories will need to have real-time feedback on how the machines are performing or if they need to make changes.

How secure could you make a CBRS system? You could have a dedicated band just for your internal use and only have it on your devices. Invisible to the outside world but giving you the bandwidth that you need in your office, warehouse, or factory.

We have seen the players be OEMs and carriers and other integrators in this space. Conspicuously absent are the cable companies. Here is space where they can shine, grow, and spread beyond Wi-Fi without building an ironclad agreement with one carrier. They have the money and the deployment process to make this a phenomenal area of growth. I would like to think that SpiderCloud would be calling the cable companies with proposals and business cases. Just my opinion. It is time for the cable companies to make it happen in wireless deployment.

Inside coverage summary

So, to sum it up, there are plenty of options that you will have. Some you have control of and some you don't. You also need to separate what the first responders will need and what other services will need. This is specific to the band and the coverage. It matters.

When planning, try not to think of just one service or area. Look at the building, then look at the service you want. Something like what's below.

1) The building has:
    a. Common areas like the entrance, the mezzanine, the food courts.
    b. Emergency exits, stairwells, basements, rooftops and other areas that are only accessed by workers, contractors, and when there's an emergency.
    c. Office space that may be secure and locked occupied by tenants.
    d. Restrooms, public, and tenant only.
    e. Hallways where people could be walking at any given time.
    f. Entrance and exit areas where people could stop to smoke on break or congregate.
2) Coverage for:
    a. Public safety should cover the entire building.
    b. Carriers will want to cover the common areas and some tenant offices.
    c. IOT coverage for the meters, air conditioners, thermostats and other controls.
    d. Private networks throughout the building in tenant spaces.
3) Emergency phone connections:
    a. Elevators
    b. Rooftops
    c. One on each floor
    d. Basement
    e. Wherever the generators are located.

Small Cell and CRAN Deployment Report

I hope this helps you with what needs to be added to the buildings in a smart city.

Small Cell and CRAN Deployment Report
The Common Carrier Small Cell

When will we have a common carrier small cell? A small cell that all the carriers could share in any given venue. It has been promised by so many, but nothing yet. Why? Well, we already have it, in a way. Simple Wi-Fi is already a common carrier, right? In a building, if you have Verizon, and your friends have T-Mobile, AT&T, and Sprint, then all your smartphones will work on the common Wi-Fi. The same could be true for LTE-U and CBRS small cells. Yet, the carriers may never let that happen. It would offload traffic, but they would lose control.

Many small cell OEMs have promised this, but it takes more than a cool device. It takes spectrum. Now we will have that in LTE-U and CBRS. So, we thought, but the CBRS was delayed since T-Mobile wanted more spectrum for a longer time. The FCC has been holding this up. The delay helped Sprint gain traction with their 2.5GHz spectrum, but I digress.

Not everyone wants to share. Verizon and AT&T appear to want to be exclusive wherever they can. They don't like to share any assets from what I have seen. I don't blame them; they can afford to pay a premium and Verizon is known for its great coverage. They want to maintain any edge they can. I get it.

The CBRS will allow the carriers to have a shared small cell using common spectrum, once the devices all get the spectrum in them. I don't think they will agree to do it, but we'll see what the future of billing models brings. It may make sense for them to do it to reach more users.

If you don't think that a business plan must align with the common carrier, think again. As flat-rate plans become commonplace, the carriers will look to offload more and more data. They won't want to invest more than they should in backhaul or infrastructure. They may give up some control to make this happen.

If you think about it, a shared small cell or mini macro will make a lot of sense for offload. They already trust Wi-Fi to offload data, why not take it one step closer. It makes sense to use it that way now that the system is all data anyway.

I think that T-Mobile has taken the first step by using Voice over Wi-Fi. They have had some success with it. It worked well where the user had dedicated Wi-Fi, and if would offload voice to a Wi-Fi system rather than their spectrum. I thought it was cool they did it. I don't see them advertise it anymore, so it probably didn't work as well as they had hoped. I think that the CBRS and LTE-U spectrum may work better for them.

Small Cell and CRAN Deployment Report

The business plan would take some refining. The issue is who would pay to install the small cell and maintain it? It comes back to a DAS issue where someone must install and maintain it. Who pays for it and what would a monthly fee be for something like this. I'll tell you now unless it's an attractive venue, the carriers won't pay a dime! Why should they? It doesn't make sense for them to manage thousands of somebody else's small cells. That's why they like Wi-Fi, someone else is managing it. The user should sign on and set it up, it will work, or it won't. So simple.

The common carrier small cell would need to be managed by someone. Maybe the venue owner or the Wi-Fi provider. It would have to be upgraded as needed and there is the cost of the backhaul. It all adds up. If an independent company does it, do they have to pay rent to be in that venue? It's not so free, is it? A business model must be built to put these things in. Some type of payback must happen, just like the Wi-Fi models at larger airports, someone should pay for it. Again, it won't be the carrier unless it is their venue.

Small Cell and CRAN Deployment Report
Mounting Small Cells in the City

With densification of cities, we often try to figure out how we will mount the devices in our cities. After all, urban areas are going to be the key focus of how to grow. Smart cities are not the only reason we will be mounting the equipment in cities. We want to bring underprivileged and underserved areas up to a high-tech region so that everyone there can become an internet business owner. I am an internet business owner; it's a great opportunity for us to make something of our ideas. Ideas can be taken from our minds to reality faster than ever before. Cities are working hard to make these things happen for everyone. The don't want favoritism; they only want to give everyone an opportunity to grow and build a business. The more successful the residents are, the more successful the city is and the more loyalty it will build for customers.

So how do we do all this great stuff? We densify networks. We offer Wi-Fi, CBRS, LTE-U, mmwave, and carrier networks where the people are. IOT will absolutely require that we have a signal as close to the device as possible, not to mention the autonomous vehicles like cars and drones. While the technology is cool, we still need the physical mount. We still need to hang a box near the people. We still need to deploy a network and a backbone. That's right; we need to install a box and a cable in the city.

Cities want it to look nice and to be quiet. That is what I have learned. How do we do that? We look at what we can mount.

City Deployment Notes

I have seen some smart city questions come my way, specifically, how can the smart city be sustainable. When I think of sustainable, I think of how to make the system pay for itself. To make it a viable and long-lasting program that doesn't rely on tax dollars just to survive. Those systems often die of financial death. So, what can we look at for income and a reduction of expenses?

It's a good idea to create a plan to add income and reduce existing expenses and make it all look nice. We want to be prepared for when the opportunity comes to us. We also should be prepared to look at the existing leases and contract because it's a good idea to know where you're starting from.

Expense Reduction:

Expense reduction is not always obvious. Using lower power LED lights makes sense if the lamps are easy to replace and have a payback of less than 1 year.

This has been an issue. However, we can see where we can save money in the long run by investing now. Expense reduction is not obvious to us because we normally we look to spend money up front, but we know we can save them money in the long run. We need to build a good long-term case because technology people look at things so differently than the financial guys do. For example, the IT people see a new server that is energy efficient and has a low heat footprint is going to save money on not only energy savings but with less heat the cooling isn't so critical saving money on the air conditioning bill. Most financial people will look at the expense to buy the new equipment. They see the one-time bill, not the big picture. That is why the technology groups need to provide a bigger picture when possible. Don't worry because most OEMs will do it for you so that they can sell their equipment. I just want you to be prepared to get the right data. To prove to the financial people that we have the answer for long-term sustainability, cost-effectiveness and that this is a long-term solution.

I know what you're thinking, I just avoided specifics. Well, here they are. The obvious, LED lighting, which is coming down in the upfront costs. Solar energy systems which are also becoming more and more cost-effective, especially where there is plenty of sun. Not so effective in the northeast just yet, but I see them all over California in newer communities. What about low energy equipment? Wi-Fi has already learned this, they lowered power and became creative with the way they connect power, this has really helped the cable companies deploy Wi-Fi everywhere. Just like their interfaces. It works well and its reliable, and it's made to be outdoors. They have Wi-Fi in street lamps, (in the lamp or in the post), so there is no obvious equipment on the post, Dallas is already doing this as is Los Angeles. That's right, it's real and being deployed. Just like LTE will be soon. (I am not talking about smart light bulbs in your home, they connect to Wi-Fi, but they are not a hot spot.)

I know that aesthetics matter in the city. So, they could put in a new pole with new lighting that takes less energy. They could also have a pole that is ready for the growth of wireless equipment inside the pole. Maybe they could think through the poles to add small cells, Wi-Fi, and routing equipment with fiber access. Then what? You make money off the asset. This is where you really have something to gain. Replacing the pole costs money, and if they must run new power for the lamps, it will be a lot of money. Older lamps ran at a higher voltage, so it could be cost inhibitive now.

Small Cell and CRAN Deployment Report

Now, let's look at ways to make money of existing city asset.

The low hanging fruit for the city is to lease the assets, like poles, lampposts, traffic lamp posts, strand, and more. Just be smart when you negotiate the lease for the poles, remember that most carriers want the pole to themselves, but don't get into an exclusivity contract for all poles. Be smart and plan it out. Also, if you don't' want to manage the poles, then work out a deal with a company that already is good at leasing poles. Make sure you get a piece of the rent for each pole. While up front, the income will seem low, as the need grows, you will have more income from new players. This is tough on many poles because there are already local telecom and cable companies there. One thing that many cities learned is that many of these poles with fiber on already have contracts in place where they limit the competition getting on new poles. Don't limit your options. Make sure you put in there that it must look nice and be quiet.

Be quiet? What does that mean? It means that most older equipment had loud fans that really bothered the residents. If it's in a residential neighborhood, then it should be quiet. It must look nice, so people don't complain.

Let's look at other income, do you already have fiber or wireless backhaul. This is something that you could position to lease or offer carrier services. If you don't want to manage it, then you could work with a fiber distribution company to manage your fiber. It's a source of income that could be utilized.

You would provide space for kiosks that would be great for Wi-Fi access and digital signage. I see two models, one where you just lease the space and let the company build and do it all. The second model is like what NYC has done, build the kiosk, put in Wi-Fi, and provide digital signage. You could make money from the rent, of the Wi-Fi access, or advertising on the digital signage. The choice is yours.

Obviously, you will make money with the permits, just be smart about it, if you put crazy restrictions in place where you should inspect each one, you will lose money. If you are very clear on what the installation should look and sound like, then you can expect a quality installation and inspect a sample. To do this efficiently, lay out the expectations up front, don't leave any room there to guess. Put in what the pole should look like, how they should run the cables underground, how they should mount the antennas, how it should be quiet, how the lamp should look, who can attach wires to it if that applies, where the fiber access should be and where the fiber equipment should be mounted (inside the pole, in a box, or underground. All the details should be thought out ahead of time, not after the fact. Be careful with whom you and out permits too. Pay attention to who does good work and who cheats.

Permitting also could be for the underground access. Remember that if someone something in the ground then you have control of it. I know many companies put in boxes underground for access and they look great. Underground boxes can add income to the city, and it can keep the equipment out of sight and out of mind. They can put antennas on manholes, it's a new frontier for Wi-Fi and small cells, to put the signal on the ground and shoot up.

Don't forget the apps! The city can provide apps to promote city business. Remember that the goal is to grow the business, so partner with then, like your local chamber of commerce, will be happy to have a way to promote business in the city. Make it easy with the apps that residents and tourists can add to their smart devices. Make it easy!

With city-owned Wi-Fi, the workers will have broadband access all over the city. Give the workers a free account and have them use it for access to databases, trouble ticket systems, timekeeping, and so on. Use the network that is there.

Use your assets to give remote access to your **parking meters,** another source of income. If people can pay with apps and credit cards over a wireless backhaul, then your revenue is not only going up, but you don't need to dump the change out of the meters. You can also track what parking spaces are busy all the time and which ones never are used. Less expense and easier income along with real-time analytics make the city a winner. If the **parking garages** are city-owned, then it's a bigger way to get income and track open spaces.

If the smart city is going to survive and grow, then put some proper planning in place. A connected city is part of the smart city. A smart city is a happy city. A smart city has the foundation to grow and attract more residents and visitors. Sustainability is key to making this happen.

City Asset Audits

If you think you have a plan for a smart city, great. If you don't them decide what to do, just don't waste this time by not knowing what you have.

One thing I have seen with Wi-Fi rollouts and Fiber rollouts is that in most cases the city is guessing at who owns what. I get it; there are a lot of poles, holes, and cables run throughout the city. I don't believe that on a person can really get a handle on who owns what.

What you can do is audit what you have. Learn who owns what and who can attach to it. This is something that will take time and spread across departments, utilities, and services. These are going to vary and will be something that the city can leverage to make income, rollout services, or make changes to save money. The audit is going to be key. If you don't think that your internal teams can do it, then hire a group to do it.

How do you do any of this if you don't know what you already have?

What assets can we mount in a city?

Strands

Think about how the cable companies utilize the strands that go from pole to pole. This way they don't need permits to attach to a pole and they don't need to pay the pole owner rent; they just attach to a strand which they put in.

Cable companies use strands to mount the Wi-Fi boxes they hang. It makes rolling out the equipment quick and easy. Fewer permitting issues, the lowest thing on the poles, so it's less of an issue to attach, put it near an access spot with a DOCSIS interface and installation is easy and quick. Easy and quick to roll out. The backhaul is generally the cable connection. The power source is also the cable connection. It's very efficient and cost-effective. Why don't cable companies roll out small cells?

Mounting assets (lampposts, wood poles, telephone poles):

The assets I am talking about could be one of many things. The obvious would be the poles, rooftops, tower, and anything that you can mount a radio or fiber. Think about all the poles and map posts and guy wire that a radio can be mounted to!

We now live in the age of constant and never-ending connectivity. Think about what we can do if we get the wireless signals out to the people! It will be a necessity at some point. The owners of the poles and lampposts in the city can really benefit. I don't think I am telling you anything new. This is already something that is being hunted down by most carriers in all cities. They want to mount radios to get the signal as close to the people as possible.

You, as a city, should know who owns what. If it's you, then great, if it's a utility, great! Just make sure you have it documented somewhere online so when someone tries to gain access they know whom to talk to and how to fill out the permits. The zoning information would be helpful too. Make it easy for them.

Don't forget what your requirements are for noise and aesthetics. These are things that a company like Crown Castle or Verizon will need to know when they deploy. If they put something up that you don't like or is noisy, then the residents complain. Usually not to them, but to the city. They talk of problems and the eyesore and of course, the noisy fans. It matters to them because they live 20 feet from it. They pay their city taxes, and they want to have a beautiful and quiet neighborhood. That's why it really helps if you do your part and define in detail what you expect prior to issuing any permits.

This is the one thing that I learned from muni Wi-Fi, the poles are poorly managed in many cities. They don't' worry about it or care about it. They let it up to the contractors, us, to figure it out. Now they may regret not keeping up with it because it takes a lot of time and walking and research to figure it all out. I wouldn't give it away for free. I would sell it, or I would keep it which means that it's of no use the next time because things may have changed. See the problem; old data could be bad data. Maybe a great place to start, but maybe starting over would be easier.

What about mounting fiber to the poles? Often, we think it's underground, but it could be overheard. The issue may be who owns the pole. If someone else signed an agreement that can refuse a competitor mounting to the pole, then you have limited assets. If you don't' think this happens, then look at http://www.tennessean.com/story/news/local/2016/10/25/comcast-sues-metro-over-google-fiber-backed-pole-otmr-ordinance/92748490/ where Comcast did all that they could to block Google Fiber. It became a court battle, https://consumerist.com/2016/09/20/comcast-att-try-again-to-stall-google-fiber-in-nashville-by-writing-law-to-slow-it-down/ where they could not get along, not at all.

So, what's a city to do? Look at the agreements you signed with your cable and fiber businesses. Look at how the utilities structured the contracts if they did so at all. Most utilities didn't care until recently. Those contracts traditionally have been a headache for them. I get it, they are a pain to manage, and that's not their primary income. However, they need to play nice in this new world of 5G!

Mounting to the pole is how most carriers do it. They like to have 360 coverage, generally 3-panel antennas but they could have 2 or one. Omni antennas still serve their purpose for the carriers, but they prefer sectors to manage traffic. The small cell and antenna will go on a pole. The fiber is the preferred backhaul for carriers. They mount an antenna or 3 on the pole; they put a box on the pole with the radio head and fiber equipment, then they are done. Prior to doing any of this, they need to get rights to the pole, sign a lease or agreement, get fiber to the pole, get the permit to mount to the pole, get permits for the fiber to the pole, and get power to the pole. There is so much prep and permitting that happens prior to any carrier getting on any pole almost everywhere. It adds to the cost. Small cell hardware is cheap. Installation is cheap. Backhaul, permitting, planning and leasing are expensive.

Underground assets:

OK, this is technically a mounting asset, but why not separate it out. Now the Wi-Fi companies are getting creative in getting the signal to the people. They are working to provide coverage even if it's on a manhole cover. Who owns the manhole? You should have that documented somewhere.

What about putting in vaults to mount the radios and router equipment so that it can connect to the fiber? Yes, underground vaults are a brilliant idea that is coming of age so that all we need to put on the pole is the antenna and maybe a very small radio head. Who will own that asset? I say the city, and they have a beautiful underground radio vault where they can charge rent.

What about the routing of cable? Here is another place the city or utility can allow access to the fiber and cable runs. This is something that they can lease. If you just want to stop the roads from getting ripped up every 2 years, plan with empty conduits so that future runs can be fed from manhole to manhole. Make it easy and clean to allow a new player to come in and run cables by getting access to conduits that the city and utility planners put in ahead of time to save the streets from getting ripped up. Plan ahead, my friends!

Small Cell and CRAN Deployment Report

There are solutions where Wi-Fi has been deployed on manholes for coverage. Fiber and small cells are put underground near the poles to keep the poles clean and pretty. Fiber is being run underground where possible to avoid those overhead and exposed cables from hanging off the poles and looking ugly. Underground can look nice, but as we grown and make changes, it's a nightmare. We need to pull tons of permits and rip up the pavement and spend a ton of money to add a few strands of fiber just, so we can grow or add new equipment.

Fiber assets:

You may have unused fiber you don't think you need that you could lease or sell to someone in need. It's extra income. If you don't need it or don't see an immediate need, sell or rent it.

You may be able to share some strands that you have with another carrier, business, or a customer. I know you may not want to get into the fiber business, but you could have a company manage this for you so that all you see is the reoccurring income.

These are assets that you may already have that you don't know about. Make the most of it and get some income if possible.

Building tops and Towers:

You probably know what towers and building tops you have, I can only imagine. However, can you lease space off them? In the past, you tried to keep it secure, but in this communication centric world, you can start opening this revenue stream.

Many times, you have more value than you originally thought. The rooftops that are empty because at one time they were too low now are closer to the public and possibly the best height for small cells to be mounted on edge. Open your perspectives and see what you should offer.

Your tower may be loaded on the top, but what about the lower parts? Are they open? Are they near busy places in the city? Take advantage of them and open them up!

Remember that the building is still a great way to mount the equipment. The buildings are not only the roof, don't limit yourself. We can use the outside walls to mount small cells. We can use the windows in stairways to get the signal out to the people. We can put small cells in the windows of large buildings with storefronts. Why not, it's easy and effective if the glass passes RF. Let's get creative here. Why not work with small businesses and give them free access to the broadband to get inside their building and get the signal on the street. It just makes sense to me. It's an asset that could be a win-win for everyone involved.

Billboards:

This is something that the city may not own, but they could have access to. Many billboards have power which means they could be prime real estate for small cells and Wi-Fi and IOT! Use them. Find out who owns the property and who manages them. Keep track of them and see if you can offer them to wireless providers coming into your city.

Kiosks, billboards, and signs – here is something that is really underutilized by most cities, they need to use city-owned assets beyond the bus stops. Billboards are obvious, they are everywhere and prove to be valuable in mounting wireless assets because most already have power. Many larger cities have kiosks to help people around the city. They have signs showing people city maps. They have pay phones that may be there for emergencies. I recommend using these assets or renting them so that broadband can get out to the city for more people to use. It could be a game changer by using something you already have available to provide new rental opportunities.

We need to get more creative to roll out new wireless formats. Whether it's Wi-Fi, 5G, IOT, or any other format, let's get the assets out there and see what we can realistically mount to.

Think about what the options could be! We can do so much more in a Dense Network. Dense networks are becoming mainstream for the world, so let's get started here in the USA.

Think they don't matter? What about your public safety systems? As FirstNet rolls out, they are going to push to mount in cities at some point. Make this another place they can use. AT&T will need more space, they want to cover your city, and you can bet they will use the FirstNet name to mount anywhere they can. Why not?

Small Cell and CRAN Deployment Report

Parking Garages:

I often see city-owned parking garages in smaller cities. This is a great place to put antennas on. The stairwells are a great place to mount Wi-Fi and small cells. Lower levels could be used for small cells. This is another thing you should look at in your audit to see what you should create a new source of income. Get the word out so that you expand your portfolio in a good way.

How will you know if you don't make it available to all?

Street Furniture:

You have bus stops; train stops, parks, benches, garbage cans, and dumpsters, maybe even kiosks that would be a great fit for small cells, Wi-Fi, interactive displays, and more. Why not use them? You have the property already in place, and if you have power to them, you're all set. The carriers may want to run fiber but ask if they could use wireless backhaul or an alternative to fiber. See if the fiber is nearby, it may not be so bad to run it there after all.

This may be a great opportunity to update your bus stops and train stops, make the most out of this. If you already had the plan to upgrade, think of what other services you could add. If they are owned by the transit company and not the city, then partner with them to improve what they have. Use LED lighting to save costs, add Wi-Fi to add value. So much more can be done. We can add small cells to get the carriers in there to improve their coverage and collect some rent along with it. As IOT coverage expands, it adds another source of income and service for the city and the transit company. It pays to have partners.

You could have large garbage cans or recycling cans, or donation centers that people go to that are big and fixed. Take advantage of those structures to add small cells, Wi-Fi, or even a kiosk to share with the city. This is a form of street furniture that has value to a wireless rollout.

Public Transportation

We will be looking at buses and trains to have Wi-Fi. They could also pass the signal outside to the people there. They could have hot bus stops and train stations that can spread the signal to the people.

Wireless Backhaul:

If you have microwave or data backhaul, it could be used and leased to others. Many times, old microwave shots will be taken down or not used, so why not put in something that could serve your city or the businesses nearby. It's worth considering. It could be Wi-Fi or a point to point link or a multipoint link. If you connect those links to the internet, you could connect buildings to a backbone. That's something that you could offer to small businesses in that building, a broadband connection.

Data Collection:

I know this is a strange thing to audit, but it's an asset that is often overlooked. Think of how this could be useful. You may not see value in the data that your team has collected, but it has great value. Local business and non-profits need this data! Residents need this data! You use it internally so often that you take it for granted, but it has value.

If you give it away or if you use it internally, it can really help more than your teams. If you share it, either sell it or give it away or a combination, then your city can grow exponentially. This is something that small businesses, partners, and residents will find useful when deciding what to do next.

Small businesses may want to expand to a new neighborhood, or they may want to start a new business. The data that you provide will help them expand with great knowledge. It helps them make an educated guess using what they know, their customer feedback, and your data.

Partners and non-profits could be trying to help the city do something new, innovative, or charity work. Why not help them streamline what needs to be done and where to start. This is all going to be taken from what they learned and the data you provide. They want to help you become better, a better city with a better reputation. Do it as a team, play the part of supporter.

Residents are looking for what to do and what neighborhood to live in. The data you provide can help them make that decision. They need your input so that they can improve their daily lives. So that they can brag to their friend and family about what a great city they live in. Help make them proud of your city!

Notes:

I just want your teams to think outside the box and see value in more than the physical assets. I often fall into this being a wireless guy. I see so many things that are useful but not taken advantage of. Why not use all of them.

Small Cell and CRAN Deployment Report

Resources:
- https://zebu.io/smart_cities/s_asset.html
- http://ibhc.com/services/utilities/inventory-assets/kansas-city-power-light-kcpl-pole-inventory/
- http://www.gislis.org/2011symposium/presentations/5c_mobile_pole_inventory.pdf

Acronyms and Definitions

- **AI** - Artificial Intelligence or Augments Intelligence.
- **CBRS** – Citizens Broadband Radio Service – in the USA this is 3550MHz to 3700MHz, often referred to as the 2.5GHz spectrum. Learn more at https://www.leverege.com/blogpost/what-is-cbrs-lte-3-5-ghz
    a. **ASA** - Authorized Shared Access
    b. **PAL** - Priority Access Licensed
    c. **LSA** - Licensed Shared Access
    d. **GAA** - General Access User
- **CCI** – Crown Castle Incorporated
- **CLEC** - Competitive Local Exchange Carrier
- **Cmwave** – spectrum in the 3 to 30 GHz range and will most likely be used in 5G for the fixed spectrum but could have mobility potential.
- **CRAN** – Centralized RAN
- **cRAN or C-RAN** – Cloud RAN
- **DAS** – Distributed Antenna Systems
- **FTTH** – Fiber to the Home
- **FTTSC** – Fiber to the Small Cell
- **FTTx** – Fiber to the Anything
- **FWA** – Fixed Wireless Access
- **IOT** – Internet of Things
- **KPI** – Key Performance Indicators
- **LoRaWAN** – Long Range Low Power WAN
- **LOS** – Line of Site
- **LPN** – Low-Power Network
- **LPWAN** – Low Power Wide Area Network

Small Cell and CRAN Deployment Report

- **LTE –** Long Term Evolution
- **LTE- U –** LTE Unlicensed, generally license-free spectrum in the 2.4GHz and the 5.8GHz ISM bands
- **MEC –** Mobile Edge Computing
- **Metro cell –** larger coverage area small cell
- **Microcell –** small coverage area small cell
- **MIMO –** Multiple in Multiple out
- **Mini Macro –** a very large small cell, smaller than a macro BTS
- **Mmwave –** spectrum in the 30 to 100GHz range that will be part of 5G and most likely used for fixed wireless.
- **NB-IOT –** Narrowband Internet of Things
- **NLOS –** Near or No Line of Site
- **NR –** New Radio format developed by QUALCOMM to be used in 5G radios
- **OTF –** Off the shelf, a term used to describe common equipment, like servers or routers.
- **PaaS –** Platform as a Service
- **Pico Cell –** small business, lightly load small cell
- **PoE –** Power over Ethernet
- **POTS -** Plain Old Telephone Service
- **PTMP –** Point to MultiPoint
- **PTP –** Point to Point
- **PTT –** Push to Talk
- **RAN –** Radio Access Network
- **ROW -** Right of Way
- **SaaS –** Software as a Service
- **SAS –** Small Cell Antenna Systems, like DAS systems but with small cells only.
- **SCaaS –** Small Cell as a Service
- **SDN –** Software Defined Networking
- **SISO –** Single in Single out (antennas)
- **UE –** User Equipment, like a smartphone
- **URLL -** Ultra Reliable Low Latency.
- **Wi-Fi –** Wireless Fidelity, generally license-free spectrum in the 2.4GHz and the 5.8GHz ISM bands

- **WiMAX** – Worldwide Interoperability for Microwave Access, based on 802.16 set of standards, learn more at https://en.wikipedia.org/wiki/WiMAX
- **VoLTE** – Voice over LTE
- **Femto Cell** – home use small cell

I know most of you think that fiber will take over, but until fiber can carry power, (spoiler alert - it can't alone!), then we need CAT5 or CAT6 or whatever else they come out with. CAT5 has been used for over 15 years, and it's still going strong. Don't deny it, we love wireless, but we need CAT5 somewhere.

Thank you Again!

Thank you for your support. I pray that it serves you well.

I want to thank you for your thirsting for technical know-how. The people that want to learn more will grow in every way. I am here to serve that need.

If you need one on one consulting or specific reports, feel free to reach out at wade@techfecta.com or wade4wireless@gmail.com for direct support.

About TechFecta

Wade is the Chief Technology Analyst for TechFecta. TechFecta, Tech consulting for the real world. www.techfecta.com. He is also a Solutions Consultant, Technology Analyst, Technology Marketer. Author, blogger, podcaster.

Blog and podcast available at www.wade4wireless.com if you want to follow.

Link up with me on LinkedIn, https://www.linkedin.com/in/wadesarver/.

Reach out to Wade on LinkedIn or at Wade@techfecta.com or wade4wireless@gmail.com to send feedback.

Twitter @wade4wireless, https://twitter.com/Wade4Wireless.

TechFecta is your CTA, Chief Technology Analyst!

WHAT WE DO:
- Tech Reports and Books
- Tech Research
- 5G Deployment Consulting
- Technology Marketing
- Investor Support
- Live Consulting
- Smart City Planning and Deployment Consultations.

WITH WORKING KNOWLEDGE OF:
- Wireless Deployment, including carriers, distributors, OEMs, and contractors.

- 5G deployment, massive MIMO.
- 5G technology.
- Wireless Technology.
- Tower Climber Industry.
- Smart City Deployments.
- Field equipment, vendors, carriers, and tower companies.

More Reports and Books:

The blog is at www.Wade4Wireless.com to offer real-world knowledge to wireless workers and investors.

TechFecta is here to help workers, businesses, and investors gain clarity on the direction technology is heading from where it is today.

- The Massive MIMO Report, https://wade4wireless.com/2018/06/03/the-massive-mimo-deployment-report/
- The Backhaul Deployment Report, https://wade4wireless.com/2018/05/28/the-mobility-backhaul-report/
- The New T-Mobile Consolidation Report, https://wade4wireless.com/2018/05/13/the-new-t-mobile-consolidation-report/
- The 5G Deployment Plan Handbook, https://wade4wireless.com/2017/01/30/the-5g-deployment-plan-book-release/
- Wireless Deployment Handbook for LTE Small Cells and DAS, https://wade4wireless.com/2015/11/12/wireless-deployment-handbook-for-lte-small-cells-and-das/
- The Smart City Tech Planning Handbook, https://wade4wireless.com/2017/08/11/the-smart-city-tech-planning-handbook/
- 1099 Worker or W2 Employee Report, https://sellfy.com/p/SCil/
- Learning 5G in the Real World, https://wade4wireless.com/2017/09/18/learning-5g-in-the-real-world/

- [SCOPE OF WORK TRAINING](http://wadesarver.com/scope-work-training/), http://wadesarver.com/scope-work-training/
- [Smart City Use Cases Report](https://wade4wireless.com/2017/11/22/smart-city-use-cases-report/), https://wade4wireless.com/2017/11/22/smart-city-use-cases-report/
- The "Secrets to 5G Deployment" book coming soon!

Figure 12 Wade

Be smart, be safe, and pay attention!

See ya!

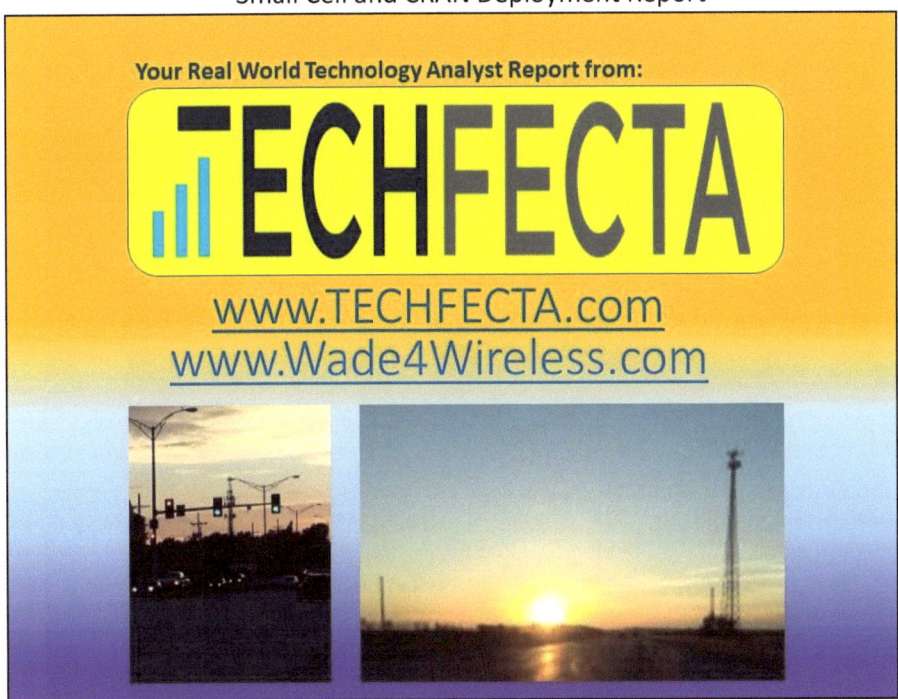

Figure 13 Back Cover

www.ingramcontent.com/pod-product-compliance
Lightning Source LLC
Chambersburg PA
CBHW040315220526
45473CB00009B/2445